Medicinal Plant Biodiversity
of Lesser Himalayas-Pakistan

Arshad Mehmood Abbasi · Mir Ajab Khan
Mushtaq Ahmad · Muhammad Zafar

Medicinal Plant Biodiversity of Lesser Himalayas-Pakistan

Springer

Arshad Mehmood Abbasi, Ph.D.
Department of Plant Sciences
Quaid-i-Azam University
Islamabad, Pakistan
arshad799@yahoo.com

Mir Ajab Khan, Ph.D.
Department of Plant Sciences
Quaid-i-Azam University
Islamabad, Pakistan
mirajab@hotmail.com

Mushtaq Ahmad, Ph.D.
Department of Plant Sciences
Quaid-i-Azam University
Islamabad, Pakistan
mushtaq@qau.edu.pk

Muhammad Zafar, Ph.D.
Department of Plant Sciences
Quaid-i-Azam University
Islamabad, Pakistan
zafar@qau.edu.pk

ISBN 978-1-4614-1574-9 e-ISBN 978-1-4614-1575-6
DOI 10.1007/978-1-4614-1575-6
Springer New York Dordrecht Heidelberg London

Library of Congress Control Number: 2011943095

Printed on acid-free paper

Springer is part of Springer Science+Business Media (www.springer.com)

Preface

Although medicinal plants have a rich history of utilization in all cultures, no one knows when or where plants first began to be used in the treatment of various ailments. From the very beginning of human existence, humans have familiarized themselves with plants and have used them in a variety of ways throughout the ages. Primitive humans began to distinguish those plants suitable for nutritional purpose from others with a definitive pharmacological action. This relationship has grown between plants and humans, and many plants have come to be used as drugs. Ethnobotany is the traditional knowledge of indigenous communities, about the surrounding plant diversity, and the study of how the people of a particular culture and region make use of indigenous plants. The Himalayan region is among the largest mountain systems of the world, with uncounted and unique medicinal plant resources. Interactions between humans and the natural system help in maintaining the richness of species, communities, and genetic materials on both productive and wild lands. The Lesser Himalaya range is an extension of the mighty Greater Himalayas. The Lesser Himalayas have a unique ecology, vegetation, and diversity of medicinal flora due to tremendous variations in the altitude, climate, and associated wildlife.

The utilization of medicinal plants in modern medicine suffers from the fact that although plants are used to cure diseases, scientific evidence in terms of modern medicine is lacking in many cases. Different societies use plants according to their own beliefs, knowledge, and previous experiences. Their knowledge about the use of the plants is usually not known to other societies or to scientists. These hidden areas need to be explored. This book provides a brief introduction to the Lesser Himalayas' ethnobotanical aspects, marketing, and anthropogenic pressure on medicinal flora. It comprises 100 medicinal plant species, including pteridophytes, gymnosperms, and angiosperms (monocots and dicots), along with their scientific descriptions and traditional uses.

We have tried to convey a maximum of knowledge regarding medicinal plant diversity in a minimum of words. There is always room for improvement. Readers are the best judges to evaluate this effort. We believe readers have a moral obligation to convey suggestions for our book's future improvement. It would be our greatest achievement if this book could attract students of botany, biodiversity, plant taxonomy, ethnobotany, and ecology, as well as wildlife naturalists, tourists, and others who have some lovely feeling for nature.

Islamabad, Pakistan

Arshad Mehmood Abbasi, Ph.D.
Mir Ajab Khan, Ph.D.
Mushtaq Ahmad, Ph.D.
Muhammad Zafar, Ph.D.

Contents

List of Figures

Chapter 1
Introduction: Lesser Himalayas—Pakistan and Medicinal Plant Wealth

Abstract The Himalaya Range is a mountain range in Asia, separating the Indian subcontinent from the Tibetan Plateau. It is the name of a massive mountain system that includes the Karakoram, the Hindu Kush, and other lesser ranges that extend out from the Pamir Knot. The Himalayan mountain system is the planet's highest mountain range and home to the world's highest peaks. The Pakistani Himalayas are located south and east of the Indus River, which originates close to the holy mountain of Kailash in western Tibet, marking the range's true western frontier. The Lesser Himalayas are a prominent range 2,000–3,000 m (6,600–9,800 ft) high formed along the Main Boundary Thrust fault zone, with a steep southern face and gentler northern slopes. These Himalayas lie north of the Sub-Himalayan Range or Siwalik Range and south of the Great Himalayas. The Lesser Himalayan range is one of the unique emporiums of medicinal plant diversity. The most commonly utilized plant species are *Berbers lyceum*, *Bergenia ciliate*, *Cassia fistula*, *Cichorium intybus*, *Datura innoxia*, *Justicia adhatoda*, *Mallotus philippensis*, *Melia azedarach*, *Mentha longifolia*, *Mentha royleana*, *Myrsine africana*, *Punica granatum*, *Solanum nigrum*, *Viola canescens*, and *Zanthoxylum armatum*.

Keywords Climate • Flora and fauna • Folk recipes • Hydrology • Lesser Himalayas—Pakistan • Medicinal plant processing

1.1 Pakistan

Pakistan stretches from the Arabian Sea to the high mountains of Central Asia and covers an area of 803,944 km^2. It lies approximately between 24° and 37° north latitude, and between 61° and 78° east longitude. It neighbors Iran to the west,

A.M. Abbasi et al., *Medicinal Plant Biodiversity of Lesser Himalayas-Pakistan*,
DOI 10.1007/978-1-4614-1575-6_1, © Springer Science+Business Media, LLC 2012

Afghanistan to the north, China to the northeast, and India to the east and southeast along a 2,000-km, partially contested border. There is a 1,000-km coastline along the Arabian Sea.

1.1.1 Physical Regions

Pakistan can be divided physiographically into four regions:

1. The great highlands
2. The Balochistan Plateau
3. The Indus Plain
4. The desert areas

The Himalayan and the trans-Himalayan mountain ranges, rising to an average elevation of more than 6,000 m and including some of the world's highest peaks, such as K2 (8,611 m) and Nanga Parbat (8,126 m), make up the great highlands that occupy the northernmost part of the country. The Balochistan Plateau, a broken highland region about 300 m in elevation with many ridges crossing it from northeast to southwest, occupies the western and southwestern sectors of the country. The Indus Plain, the most prosperous agricultural region of Pakistan, covers an area of 520,000 km^2 in the east and extends to 1,100 km from northern Pakistan southward to the Arabian Sea. In the southeast are the desert areas.

1.1.2 Biodiversity

Pakistan has 225 Protected Areas (PAs), 14 national parks, 99 wildlife sanctuaries, and 96 game reserves. It is a land of rapidly shrinking wetlands, some of them of international significance, of wondrous juniper forests, of minute life forms that buzz their way to a magical existence, of stunning mountains, and much more. Pakistan covers a number of the world's ecoregions, ranging from the mangrove forests stretching from the Arabian Sea to the towering mountains of the western Himalayas, Hindu Kush and Karakoram. The country lies at the western end of the South Asian subcontinent, and its flora and fauna are composed of a blend of Palearctic and Indomalayan elements, with some groups also containing forms from the Ethiopian region.

1.1.3 Ecological Zones

Pakistan is divided into nine major ecological zones. The World Wildlife Fund–Pakistan (WWF-P) is working to conserve the environment through its Target Driven Programmes (TDPs), which address issues pertaining to samples of forest, freshwater, marine ecosystems, species, toxics, and climate change. The emphasis is on conserving representative sites of ecologically important areas within these Target Driven Programmes. Conservation of desert ecosystems is included under forests. In most of its projects, WWF-P supports local community initiatives to conserve natural resources and helps look for ways to improve community livelihoods. Almost all conservation projects have the following common features and priorities: partnership with local bodies and capacity building at all levels from local communities to government bodies.

1.1.4 Critical Ecosystems

Under the WWF's Global 200 initiative, ecosystems have been ranked to carry out conservation through comparative analyses. It covers all habitats on land masses and in the oceans. The Earth has been divided into 238 ecoregions by the United Nations, the National Geographic Society, and the WWF. Five of these ecosystems are in Pakistan. The Global ecoregions of Pakistan are

1. The Rann of Kunth grasslands
2. The Tibetan Plateau
3. The Western Himalayan Temperate Forests
4. The Indus Delta ecosystem
5. The Arabian Sea

1.2 Himalayan Range

The Himalaya Range, or the Himalayas, is a mountain range in Asia separating the Indian subcontinent from the Tibetan Plateau. It is also the name of a massive mountain system that includes the Karakoram, the Hindu Kush, and other, lesser, ranges that extend out from the Pamir Knot. The Himalayan mountain system is the planet's highest and is home to the world's highest peaks, the eight-thousanders, which include Mount Everest and K2. The main Himalaya range runs, west to east, from the Indus river valley to the Brahmaputra river valley, forming an arc 2,400 km long, which varies in width from 400 km in the western Kashmir–Xinjiang region to 150 km in the eastern Tibet–Arunachal Pradesh region.

The Pakistani Himalayas are located south and east of the Indus River, which originates close to the holy mountain of Kailash in western Tibet, marking the range's true western frontier. The river enters Pakistan from India, flowing northwest to Skardu. It then continues on this bearing and is joined by the Hunza River south of Gilgit. From here it flows south and west, eventually flowing from the Himalayan foothills onto the Indian planes. The Himalayas are a totally separate range from the Karakoram, which run parallel to the north. The Himalayas in Pakistan are green and fertile as compared

Fig. 1.1 A panoramic view of Pakistan Himalayas

to the arid Karakoram and Hindu Kush further north. The Himalayas have a considerably higher precipitation level during the monsoon months, creating an environment for rich pine forests and grassy meadows that more closely resemble Canada or Kyrgyzstan than the Karakoram Mountains. The Himalayas are spread across three of Pakistan's provinces. The northern area encompasses the Nanga Parbat massif and its surrounding valleys, Azad Jammu, and Kashmir. The extreme southeast corner of the North-West Frontier Province (Pakhtoonkhawa) includes portions of the Lesser Himalayas, also known as the Middle Himalayas. As with the rest of the country, the region has a strong Muslim identity. Most residents are Sunnis, with some Shi'as in the Astor Valley's upper tributaries. Languages spoken include Shina, Pashto, Hindko, and Kohistani; however, nearly everybody also speaks Urdu. The region south of the Gilgit represents diverse ethnicities and cultures, making it interesting to visit for this reason alone (Fig. 1.1).

The Lesser Himalayas are a prominent range 2,000–3,000 m (6,600–9,800 ft) high formed along the Main Boundary Thrust fault zone, with a steep southern face and gentler northern slopes. These Himalayas lie north of the Sub-Himalayan Range or Siwalik Range and south of the Great Himalayas. They are nearly continuous except for river gorges, where rivers from the north gather like candelabra in a handful of places to break through the range. In Pakistan these mountains lie just north of Rawalpindi district, covering the districts of Batagram, Mansehra, and Abbottabad as well as Pakistan Administered Kashmir. These mountains are also home to Pakistan's important hill stations, including Murree, Ghora Gali, and Nathia Gali. It snows during a few months of the year, but no glaciers are found in this region.

1.2.1 Topography

The Himalayas are among the youngest mountain ranges on the planet and consist mostly of uplifted sedimentary and metamorphic rock. According to the modern theory of plate tectonics, their formation is a result of a continental collision or progeny along the convergent boundary between the Indo-Australian Plate and the Eurasian Plate. Such a formation is referred to as a "fold mountain." The Pakistani Himalayas were formed as a result of the collision of the Indian subcontinent with Asia. This process of plate tectonics is ongoing, and the gradual northward drift of the Indian subcontinent still causes earthquakes. Nanga Parbat stands on the southern tectonic plate, while Rakaposhi stands on the northern plate with the Indus River dividing the two. While the Himalayas formed relatively slowly, the Karakoram were rapidly pushed upwards, resulting in their comparatively dense topography.

1.2.2 Glaciers and Rivers

The Himalayan range encompasses about 15,000 glaciers, which store about 12,000 km^3 of freshwater. The 70-km-long Siachen Glacier at the India–Pakistan border is the second-longest glacier in the world outside the polar region. Some of the other more famous glaciers include the Gangotri and Yamunotri (Uttarakhand), Nubra, Biafo, and Baltoro (Karakoram region), Zemu (Sikkim), and Khumbu glaciers (Mount Everest region). Some of the world's major rivers, including the Ganges, Indus, Brahmaputra, Yangtze, Mekong, Salween, Red River (Asia), Xunjiang, Chao Phraya, Irrawaddy River, Amu Darya, Syr Darya, Tarim River, and Yellow River, rise in the Himalayas. Their combined drainage basin is home to some 3 billion people in countries including Afghanistan, Bangladesh, Bhutan, People's Republic of China, India, Nepal, Burma, Cambodia, Tajikistan, Uzbekistan, Turkmenistan, Kazakhstan, Kyrgyzstan, Thailand, Laos, Vietnam, Malaysia, and Pakistan (Figs. 1.2 and 1.3).

1.2.3 Climate

The climate of the Himalayas ranges from tropical at the base of the mountains to permanent ice and snow at the highest elevations. The amount of yearly rainfall increases from west to east along the front of the range. This diversity of climate,

Fig. 1.2 Amazing Saiful Malook Lake

Fig. 1.3 Hydrology and *Pinus* vegetation

altitude, rainfall, and soil conditions generates a variety of distinct plant and animal communities. On the Indo-Gangetic plain at the base of the mountains, an alluvial plain drained by the Indus and Ganga-Brahmaputra river systems, vegetation varies from west to east with rainfall. The xeric Northwestern thorn scrub forests occupy the plains of Pakistan and the Indian Punjab. Further east lie the Upper Gangetic plains moist deciduous forests of Uttarakhand and Uttar Pradesh and the Lower Gangetic plains moist deciduous forests of Bihar and West Bengal. These are monsoon forests, with drought-deciduous trees that lose their leaves during the dry season.

1.2.4 Flora

The flora of the Himalayas varies with climate, rainfall, altitude, and soils and includes elements from tropical Indochina, temperate East Asia, the Palaearctic region, the Deccan Plateau and the low-lying areas along with the support of mixed evergreen forests. Although most of these semi-evergreen forests have long since been converted into human uses, vestigial patches appear mostly in small protected areas. The alluvial grasslands and savannas along the foothill valleys are among the tallest in the world. Characteristic species in these highly productive grasslands include *Saccharum spontaneum*, *Phragmitis kharka*, *Arundo donax*, *Imperata cylindrica*, *Erianthus ravennae*, *Andropogon* spp., and *Aristida ascensionis*. The lower hill slopes above 1,000 m are cooler and less drought-stressed. These areas are dominated by subtropical evergreen broadleaf forests. The eastern Himalayas' temperate forests are dominated by evergreen broadleaf trees and mixed conifers (e.g., *Quercus*, *Lauraceae*, *Tsuga*, *Taxus*) in the lower reaches and winter-deciduous broadleaf species (e.g., *Acer*, *Betula*, *Magnolia*) in the upper reaches. The drier, south-facing slopes support extensive stands of arboreal *Rhododendron* species that may co-occur with oak (*Quercus semecarpifolia*) or other ericaceous species such as *Lyonia ovalifolia*. These temperate forests support a rich epiphytic community, consisting of a variety of dicots, orchids, ferns, and mosses. Bamboo (*Arundinaria* spp.) is dominant in the unexploited places, especially where it provides early-successional ground cover following fire. Subalpine conifer forests begin from about 3,000 m and extend to 4,000 m. In the eastern Himalayas, *Tsuga*, *Picea*, or *Larix* dominate these forests between 3,000 and 3,500 m and *Abies* dominates above 3,500 m. Junipers are widespread along the timberline and may form dwarf krummoltz formations above 4,700 m. The dry slopes and inner valleys support *Pinus* and *Cupressus* on basic limestone soils. Above the tree line, the vegetation is a moist alpine scrub community of dense juniper and *Rhododendron* shrubberies that extend to about 4,500 m. From 4,500 to 4,700 m, the vegetation consists of alpine meadows with a diverse assemblage of alpine herbs and smaller-stature woody shrubs, such as a variety of dwarf *Rhododendrons*, and numerous alpine herbs such as *Potentilla*, *Ranunculus*, and the alpine *Saussure* species. Periglacial and subnival communities occur in the high alpine areas above 4,700 m, where the short growing

Fig. 1.4 Floral diversity

season, high winds, and unstable soils allow only specialized plants to survive. Some of the common genera found here are *Androsace*, *Arenaria*, *Saxifraga*, *Meconopsis*, and *Primula* (Fig. 1.4).

1.2.5 Fauna

Knowledge of the fauna of the eastern Himalayas region is poor. Most of the information available is on the larger vertebrates that are easily observed and inventoried. Overall, more than 175 species of mammals and in excess of 500 species of birds are known from the region. The mammalian fauna in the lowlands is typically Indo-Malayan, consisting of langurs (*Semenopithicus* spp.), golden langur (*Trachypithecus geei*), pygmy hog (*Sus salvinus*), hispid hare (*Caprolagus hispidus*), flying squirrel (*Biswamoyopterus biswasi*), wild dogs (*Cuon alpinus*), sloth bear (*Melursus ursinus*), gaur, and several species of deer such as muntjacs (*Muntiacus muntjak*) and sambar (*Cervus unicolor*), snow leopards, Asiatic black bear (*Ursus thibetinus*), blue sheep (*Pseudois nayur*), takin (*Budorcas taxicolor*), Himalayan tahr (*Hemitragus jemlahicus*), and red panda (*Ailurus fulgens*). Some species of birds restricted to the region include the Manipur bush quail (*Perdicula manipurensis*), chestnut-breasted partridge (*Arborophila mandelli*), Blyth's tragopan (*Tragopan blythii*), Temminck's tragopan (*Tragopan temminckii*), Sclater's monal (*Lophophorus sclateri*), Tibetan eared pheasant (*Crossoptilon harmani*), rusty-bellied shortwing (*Brachypteryx hyperythra*), white-winged duck (*Cairina scutulata*), white-bellied heron (*Ardea insignis*), black-necked stork (*Grus nigricollis*), and Bengal florican (*Houbaropsis bengalensis*) (Fig. 1.5).

Fig. 1.5 Fauna and flora

1.2.6 Livestock

Cattle, goats, sheep, buffalo, camels, ass, horses, mules, and domestic poultry are common livestock of the range.

1.2.7 Culture

Although the young generation is fond of modern culture, religious bonds are very strong and the majority of the population prefers Islamic traditions and follows the path of Allah. Islamic traditions stand out in the area. People generally wear *shalwar qameez* with a blanket or coarse chador during winter season. Chapple and shoes are common footwear. The use of ornaments among females is also common in the area. Women adorn themselves with earrings and bangles.

1.2.8 Occupations

The majority of the population depends on agriculture for their subsistence, but the income from agriculture is too meager to meet the population's needs. Many local residents have left the country in order to earn more money, and others have migrated to different parts of the country. In addition to agriculture and business, all sorts of occupations are available here, including cobbler, blacksmith, weaver, carpenter, barber, and day laborer. The people do government/military work on the top of their independent private jobs.

1.3 Medicinal Plants

The Himalayan region is the largest mountain system in the world, with uncounted and unique world resources. The interaction between the mountain people and the natural system throughout history has helped in maintaining the richness of species, communities, and genetic materials in both the productive systems and wild lands of the mountain environment. However, the rich biodiversity of this region is being disastrously depleted due to human action in the last few decades. Understanding the indigenous knowledge of mountain people in relation to biodiversity resource management is one of the key issues for sustainable development of the Himalayan region today [1].

Plants, the ultimate producers, are the most important part of the world, and all other living organisms are completely dependent upon them to live. Plants are the basic source of food, which they form by using sunlight and converting it into chemical energy. This energy is used by every living organism for its life cycle, internal metabolism, and movement as well as in its combat with environmental conditions and also with other living organisms. Green plants are the only bridge between all populations of the earth and solar energy. We use this energy indirectly in the form of animal milk, meat, leather, etc. Plants not only produce food but also serve man and all other animals in so many ways. They provide a natural habitat in which wild animals can live and reproduce [2]. Herbal medicines have had a distinct position of respect from the primitive period to the present day. The practice of ethnobotanical pharmacology is as old as mankind. In Indo-Pak the first records of plant medicine were compiled in Rig Veda between 4500 and 1600 B.C. and in Ayurveda between 2500 and 600 B.C. This system traces its origin to Greek medicine, which was adopted by Arabs and then spread to India and Europe [1]. The long tradition of herbal medicine continues to the present day in China, India, and many countries in Africa and South America. In many village markets, medicinal herbs are sold alongside vegetables and other wares. The World Health Organization (WHO) has emphasized the importance of the traditional indigenous medicines, since a large majority of rural people in developing countries still use these medicines as the first defense in health care [3]. Globally, about 85% of all medications for primary health care are derived from plants [4].

Plant-based medicines enjoy a respectable position today, especially in developing countries, where modern health service is limited. Indigenous remedies that are effective, safe, and inexpensive are gaining popularity among the people of both rural and urban areas. Information from ethnic groups or indigenous traditional medicine has played a vital role in the discovery of novel products from plants as chemotherapeutic agents [5]. People living in tribal localities and villages have used indigenous plants as medicines for thousands of years because this knowledge passes from generation to generation, based on experience [6].

Medicinal plants play an important role in the health care of people around the world, especially in developing countries. Until the advent of modern medicine, humans depended on plants for treating human and livestock diseases. Human societies throughout the world have accumulated a vast indigenous knowledge over centuries of the medicinal uses of plants and related uses, including as poison for fish and hunting, for purifying water, and for controlling pests and diseases of crops and livestock. About 80% of the population of most developing countries still use traditional medicines derived from plants to treat human diseases [7]. China, Cuba, India, Sri Lanka, Thailand, and a few other countries have endorsed the official use of traditional systems of medicine in their healthcare programs. For example, the Indian systems of medicine "Ayurveda," "Sidha," and "Unani" entirely—and homeopathy to some extent—depend on plant materials or their derivatives for treating human ailments [8].

About 12.5% of the 422,000 plant species documented worldwide are reported to have medicinal value. The proportion of medicinal plants to the total documented species in different countries varies from 4.4% to 20% [9]. About 25% of drugs in modern pharmacopoeia are derived from plants, and the rest are synthetic analogs built on prototype compounds isolated from plants. Up to 60% of the drugs prescribed in Eastern Europe consist of unmodified or slightly altered higher plant products. These drugs have important therapeutic properties, including contraceptives, steroids, and muscle relaxants for anesthesia and abdominal surgery (all made from the wild yam, *Dioscorea villosa*), quinine and artemisinin against malaria, digitalis derivatives for heart failure, and the anticancer drugs vinblastin, etoposide, and taxol. These compounds cannot be synthesized cost-effectively, which means that their production requires reliable supplies of plant material. The global importance of medicinal plant materials is evident from the huge volume of trade at the national and international levels. During the 1990s, the reported annual international importation of medicinal plants for pharmaceutical use amounted on average to 350,000 megatons valued at over 1 billion USD. A few countries dominate the international trade with over 80% of the global import and export. China and India are the world's leading producing nations, whereas Japan and Korea are the main consumers of medicinal plants. Hong Kong, the United States, and Germany stand out as important trade centers. It is estimated that currently the total number of medicinal plants in international trade is around 2,500 species worldwide [9].

The traditional system of medicine in Pakistan dates back to the Indus civilization, as verified by excavation conducted in the ancient cities of Mohenjadro and Harappa as well as of the University of Taxila, which flourished during the Gandahara period. These findings clearly reveal the importance of medicinal plants in the life and religious teachings of these civilizations. Modern medicine traces its origin back to the Greeks. Greek medicine was taken over by the Romans and then by the Arabs. After being enriched with Chinese and Indian medicine, it was taken over by modern Europeans. The Muslim rulers introduced it into India and incorporated in it the native Ayurvedic medicine. This mixture is now known as Unani medicine or, broadly speaking, Eastern medicine. The traditional Indian system of medicine, known as Ayruveda, which evolved during the period commencing from around 2500 B.C., was codified and documented in 600 B.C. Ayurveda came to be associated with the Hindus, while the Muslims of the subcontinent adopted a different traditional system known as Unani. Later on, both the "Ayurveda" and "Unani" systems benefited from and complemented each other. The dominant traditional system in Pakistan is the Unani system.

In Pakistan there are almost 50,000 herbalists spread all over the country. They run their clinics in rural and urban areas using medicinal plants in crude form. According to the Unani system, Pakistan has a rich flora in which 2,000 plant species are used at one time or another, in one culture or another, for medicinal purposes, but of these 2,000 Pakistani species, about 400–600 are documented and studied for medicinal purposes. Besides local practictioners, the rural area dwellers, especially the itinerant, use the plants based on their own experience and ancestral prescription. In the Lesser Himalayas, local inhabitants use different plant species in order to cure various ailments in daily life, including *Allium cepa*, *A. sativum*, *Aloe vera*, *Acacia modesta*, *A. nilotica*, *Achyranthes aspera*, *Ajuga bracteosa*, *Berberis lyceum*, *Bergenia ciliate*, *Calotropis procera*, *Cassia fistula*, *Cedrela toona*, *Chenopodium ambrosioides*, *Cichorium intybus*, *Cuscuta reflexa*, *Datura innoxia*, *Fumaria indica*, *Justicia adhatoda*, *Mallotus philippensis*, *Melia azedarach*, *Mentha longifolia*, *M. royleana*, *Myrsine africana*, *Otostegia limbata*, *Plantago lanceolata*, *P. major*, *Punica granatum*, *Quercus incana*, *Solanum nigrum*, *S. surratense*, Burm. *Taraxacum officinale*, *Viola canescens*, *Withania somnifera*, *Woodfordia fruticosa*, and *Zanthoxylum armatum*. Local inhabitants use these plants, as 80% of Pakistan's population lives in rural areas where these plants are easily available, the prices of the allopathic drugs are out of the reach of the poor, the allopathic drug shops are scarce, and the most important factor is that people are becoming aware of the harmful effects of artificial commodities and are realizing the benefits of a more natural way of life [10].

1.4 Folk Recipes

Just like allopathic and homeopathic medicine, the traditional herbal system uses a special combination of plants to treat diseases. Different plants have different chemicals that can have different results on different organisms. The proper quantity and quality of these plants must be consumed in order to get positive results; otherwise, either no results or some mishap can occur. That is why local people are always in search of authentic recipes. Today a lot of people are using the traditional system; for example, in China, traditional Chinese medicine is relied upon for nontoxicity, and most Chinese people avoid allopathic or other systems. In the Lesser Himalayas, local inhabitants use medicinal plants in different ways, s in the forms of decoctions, extracts, powders, pastes, or juices. The mode of application for these plants is topical as well as oral. Different recipes used by the indigenous people of the Lesser Himalayas have been developed from time immemorial through experience gained during the treatment of patients afflicted with different diseases. This traditional knowledge has been transferred from one generation to the next. Traditional herbal professional healers have special names that vary by region. From a business point of view, these healers are reluctant in sharing information about their recipes because these recipes are their trade secrets. Nowadays phytochemical laboratories have standard procedures to use plant extracts directly on a wide range of bacteria, fungi, and viruses. These and other research activities on animals are also adding to the knowledge base of herbal healing.

1.5 Diagnosis and Prescription

The diagnosis of illness is commonly done through a pulse reading. The color of the patient's tongue, the patient's general physical conditions, and the color of the eyes are also used as indicators to determine the problem.

The concept of *sard* (cold) and *garmi* (hot) is a common feature to explain the patient's illness. This idea shows that a healthy person is balanced in between *sard* and *garmi* and illnesses have the ability to take on *sard* or *garmi* forms in an ill person, which are described through pulse reading and other diagnostic techniques. Prescriptions follow a balancing effect: *sard* medicines and diet for *garmi* conditions and *garmi* medicines and diet for *sard* conditions. Change in diet, specific diets, and other social prescriptions are given along with medicinal plant remedies.

1.6 Processing of Medicinal Plants

1.6.1 Collection and Identification

Area medicinal plants are collected by local inhabitants, village grocers, and local practitioners. People collect medicinal plants on the basis of their traditional knowledge and not through scientific knowledge. Most of them have insufficient knowledge about the proper time of collection, which is essential not only to maximize active ingredients, but also from the viewpoint of sustainability of resources (Fig. 1.6).

Fig. 1.6 Collection of medicinal plants

1.6.2 Preservation

After collection, the plants are dried in sunlight and shade. Mostly flowers and leaves are dried in shade, while bark, fruits, roots, and seeds are dried in sunlight for 4–8 days. Old men and women do the drying by spreading the plants on a cloth or plastic sheets on the ground. Some plants are also used fresh. The process of drying is generally crude because dust and other foreign materials get mixed in with the specimen being dried.

1.6.3 Storage

Generally, the plant material is not stored. But some plants with high medicinal values are stored in cloth sacks and plastic or glass bottles for further use.

1.6.4 Marketing

Medicinal plants are used not only by local practitioners (*hakims*), as household remedies, and by *dawakhanas* (Herbal drugs stores), but also by Pakistan's pharmaceutical industry. The business is in the hands of a few large trading houses in the areas that neither are organized nor work on scientific lines for the collection, drying, cleaning, washing, storage, and standardization of medicinal plants. Medicinal plants are either dried for further use or sold directly to the local grocers in fresh form, where the grocer does the drying him- or herself. The village grocers have to store small quantities for a short time until they are able to sell them to wholesalers of the local markets of Rawalpindi, Abbottabad, Murree, and Haripur. The fresh drug in the market is dried by spreading it in sunlight for 4–6 days, and then it is graded (pure, mixed), packed, and stored in bags ranging in quantity from a few kilograms to mounds (1 mound = 40 kg) or (1 mound = 88.1849 pounds), depending upon the mass and availability of the drug. The process of drying is generally crude because dust and other foreign materials get mixed in with them. Storage is not done hygienically and the crude drug often gets infected with insects and fungi. This results in deterioration of the dried crude drugs and ultimately causes financial loss to the traders. In order to maintain quality, storage facilities need definite improvements.

References

1. Pei SJ (1994) Indigenous knowledge of mountain people and conservation of biodiversity in the mountain ecosystems. ICIMOD, Nepal, pp 154–165
2. Latif A, Shinwari ZK (2005) Sustainable market development for non timber forest products in Pakistan. Ethnobot Leaflets

3. Goleniowski ME, Bongiovanni GA, Bongiovanni L, Palacio CO, Cantero JJ (2006) Medicinal plants from the "Sierra de Comechingones", Argentina. J Ethnopharmacol 107:324–341
4. Farnsworth NR (1988) Screening plants for new medicines. In: Wilson EO, Peter FM (eds) Biodiversity. National Academy Press, Washington, DC, pp 83–97
5. Katewa SS, Chaudhary BL, Jain A (2004) Folk herbal medicines from tribal area of Rajasthan, India. J Ethnopharmacol 92:41–46
6. Shinwari M I, Khan MA (1998) Ethnobotany of Margalla Hills National Park, Islamabad. Department of Biological Sciences Quaid-e-Azam university, Islamabad. PASTIC Printing Press, pp 1–150
7. De Silva T (1997) Industrial utilization of medicinal plants in developing countries. In: Bodeker G, Bhat KKS, Burley J, Vantomme P (eds) Medicinal plants for forest conservation and health-care, Non-Wood Forest Products No. 11. FAO, Rome, pp 38–48
8. Prajapati ND, Purohit SS, Sharma AK, Kumar T (2003) A handbook of medicinal plants. Agribios, India, p 553
9. Schippmann U, Leaman DJ, Cunningham AB (2002) Impact of cultivation and gathering of medicinal plants on biodiversity: global trends and issues. In: Biodiversity and the ecosystem approach in agriculture, forestry and fisheries, Ninth regular session of the commission on genetic resources for food and agriculture. FAO, Rome, pp 1–21
10. Said M (1970) Hamdard pharmacopia of eastern medicine. Hamdard National Foundation of Pakistan, Hamdard Academy, Pakistan

Chapter 2
Ethnobotanical Aspects

Abstract The term "ethnobotany" was first coined in 1896 by the American botanist John Harshberger as the study of plants used by primitive and aboriginal people. Since then it has been defined as the traditional knowledge of indigenous communities about the surrounding plant diversity and the study of how the people of a particular culture and region make use of indigenous plants. Indigenous communities of the Lesser Himalayas are heavily dependent on plant resources as medicines, food, fuel, fodder, and tools in beekeeping. But at the same time, due to the overuse of these plant resources, certain species are threatened. Some common threats to the floral diversity of the region include forest fire, grazing and browsing, tree cutting, climatic fluctuation, earthquake, and flooding. In order to avoid further loss of endangered, endemic, and rare species, conservation methods should be practiced as part of a long-term conservation program. Reforestation trends have been lacking among the local communities; along with regeneration activities, an alternate source should be provided to reduce the pressure on flora.

Keywords Agriculture implements • Anthropogenic pressure • Beekeeping • Ethnobotany • Fodder • Food • Fuelwood • Marketing • Medicinal plants • Timber wood

2.1 Ethnobotany

The term "ethnobotany" was first coined in 1896 by the American botanist John Harshberger as the study of plants used by primitive and aboriginal people. Since then it has been defined as the traditional knowledge of indigenous communities of the surrounding plant diversity and the study of how the people of a particular culture and region make use of indigenous plants. Ethnobotany has its roots in botany. Botany, in turn, originated in part from an interest in finding plants to help fight illness. Ethnobotany is the most important approach to study the natural resource management of indigenous people. The issues of economic compensation and protected

A.M. Abbasi et al., *Medicinal Plant Biodiversity of Lesser Himalayas-Pakistan*,
DOI 10.1007/978-1-4614-1575-6_2, © Springer Science+Business Media, LLC 2012

areas raise the problem of the divergence between conservation managers and village communities in their perception, mode of presentation, and system of resource appropriation and allocation. Conservation managers' recognition of the knowledge and practices of indigenous people would help reduce tension and conflict between these two parties [1]. Ethnobotany includes all types of relationships between people and plants. The definition of ethnobotany can be summed up in four words: people, plants, interactions, and uses.

"Ethnobotany is the study of how the people of a particular culture and region make use of indigenous plants," while the ethnobotanist explores how plants are used as food, shelter, medicine, and clothing, for hunting, and in religious ceremonies. It is the science studying "the relationship between a given society and its environment and in particular the plant world" [2].

The Swede naturalist Carl Linnaeus actually invented ethnobotany as a student during his journey in 1732 to Lapland. On July 4, 1732, Linnaeus recorded in his diary some medical remedies used by the Sami people: "I here made the following observations relative to the remedies used by the Laplanders. Their Moxa, as the Japanese call it, but which they term Toule, is made of a fine fungus found on the birch. They apply a piece as large as a pea upon the afflicted part" [3]. He published the *Flora lapponica* in 1737, which included a discussion of the ways in which specific plants were utilized by the Laplander (Sami) people.

Harshberger's definition and vision from 1896 still provide the core for the science of ethnobotany. A slight change in emphasis can be seen through a review of Cotton's current definition: "Ethnobotany is considered to encompass all studies which concern the mutual relationship between plants and traditional peoples" [4]. The early definitions of ethnobotany restricted the field to the study of how aboriginal people used plants. Botanists, explorers, and other people who traveled the globe would see a plant and then identify, classify, and name it for the purposes of science. They would ask a local resident to give the name of the plant in the local language or to specify the local uses of the plant. This resulted in numerous monographs on the cultural group uses of plants. The particular focus of such monographs would vary depending upon the specific interest of the person undertaking the study. These early attempts of ethnobotany are considered the articulation of colonial economies, imaginations, and projects [5]. They can also be seen as the basic data-gathering stage of the ethnobotanical discipline. At present, ethnobotany has shifted its focus from people's use of plants to the relationship between people and plants, which includes use, cognition, and ecology. Recent definitions of ethnobotany [4, 6–8] demonstrate a consensus on the move to include more than just use by focusing on the relationship between people and plants. However, there is not a consensus on whether the discipline should focus on all people or on traditional and indigenous peoples [4, 6]. It is evident that people who have lived in one locality for a long time have particularly rich sets of knowledge about and cognition of plants and local ecology. A more fundamental issue relative to knowledge, however, is found within the discussion of the relationship between knowledge as practice and knowledge as heritage.

Ethnobotany has its roots in botany, the study of plants. Botany, in turn, originated in part from an interest in finding plants to help fight illness. In fact, medicine and botany have always had close ties. Many of today's drugs have been derived from plant sources. However, as modern medicine and drug research advanced, chemically synthesized drugs replaced plants as the source of most medicinal agents in industrialized countries. Although research in plant sources has continued and plants are still used as the basis for drug development, the dominant interest and resulting research funding have shifted to the laboratory. To document the secret uses of the plants, ethnobotany has become an important part of our world. Recent ethnobotanical surveys among tribal populations have brought new information to the forefront, which can be utilized to improve the economy of the tribes by organizing the systematic collection of forest products and locating cottage industries, especially of herbal drugs. In China, ethnobotany was introduced as a science in the late 1970s, but deep-rooted ethnobotanical knowledge in Chinese culture can be traced back to ancient times in Chinese history. This is evidenced by the vast literature on Chinese Materia Medica and Chinese works of agriculture and horticulture [9]. In India and Pakistan, three traditional systems of medicine, namely, Ayurveda, Siddha, and Unani, are distinguished. Ethnomedicine is an area of research that deals with medicines derived from plants, animals, or minerals and used in the treatment of various diseases and ailments based on indigenous pharmacopoeia, folklore, and herbal charm. Ethnobotany has an important role in the conservation of nature, culture, and, in particular, the biological diversity and the diversity of traditional human cultures in the world. In fact, conservation and biodiversity are linked with each other. Traditional knowledge systems are hundreds or even thousands of years old and involve not only the knowledge of plants for medicine and food but also strategies for the sustainable utilization of plant resources. In these respects, ethnobotany has played a vital role in describing traditional knowledge about medicinal uses of plants and will continue to do so in the future. To discover the practical potential of native plants, an ethnobotanist must be knowledgeable in the study of plants themselves and must also understand and be sensitive to the dynamics of how cultures work. Ethnobotanists have helped us to understand the frightening implications the loss of the rain forests would bring, not only the consequent lost knowledge about tropical plants, but also the damage brought about by the loss of native cultures in their entirety, as well as the damage to the Earth's ecological health. Out of necessity, ethnobotany is a multidisciplinary science. This multidisciplinary approach gives ethnobotanists more insight into the management of forest reserves in a period of tremendous environmental stress. Unfortunately, due to human factors that have influenced the ecological balance of these delicate ecosystems, we are presently faced with the possibility of losing our forests. Ethnobotany as an emerging science has a vital role in the improvement of plants and plant products. It certainly adds to conservation and can also be utilized for value addition.

The two fundamental strengths of applied ethnobotany are

To allow the knowledge, wisdom, and practices of local people to play fuller roles in identifying and finding solutions to issues of conservation and sustainable development.

Local people are fundamentally involved in all stages of research and practical follow-up, so there is a better chance of "buy-in" and more robust solutions [10].

2.2 Medicinal Uses

Local people mostly depend on medicinal plants because these plants are good sources of materials needed in primary health care. Local people use these medicinal plants in different situations:

1. People use medicinal plants for the treatment of various ailments on the basis of indigenous knowledge passed to them generation after generation.
2. They use medicinal plants on the advice of elders, such as wise men, herbalists, and traditional practitioners.

Medicinal plants are used in different ways and situations; for example:

1. People use them singly in their own preparations for the treatment of various ailments on the basis of indigenous knowledge passed to them generation after generation.
2. They are used on the advice of elders, wise men, and religious teachers.
3. They are used with the advice of nonqualified but professional traditional medicine workers (*hakims*), who have also gained some experience through apprenticeship with some registered practitioners.
4. They are prescribed by qualified registered practitioners (traditional herbalists) of the Unani system of medicine for a wide range of diseases and ailments.

In the study areas, people collect plants based on traditional knowledge rather than scientific knowledge. Most of them have insufficient knowledge about the proper time of collection, which is essential not only for the maximization of active ingredients, but also from the viewpoint of the sustainability of resources. After collection, the plants are dried in sunlight and in shade. Mostly flowers and leaves are dried in shade, while bark, fruits, roots, and seeds are dried in sunlight for 4–8 days. Old men and women do the drying by spreading the plants on a cloth or plastic sheets on the ground. Some plants are also used in fresh condition. Generally, the plant material is not stored. Today's younger generation in the area is forgetting the indigenous knowledge of plants; with increasing labor costs and people's search for better job opportunities, the plant collection is rapidly declining.

Among the plant parts, leaves, aerial parts, fruits, bark, flowers, rhizomes, roots, tubers, rinds, seeds, and bulbs are commonly used. All medications are classified into two types:

1. Single-plant-based
2. Based on more than one plant species

In the majority of cases, these medications were prepared by using water as a medium and administrated along with milk, ghee, oil, egg, sulfur, and butter. The method of preparation falls into different categories, including plant parts used in crushed form, in powders, in pastes, fresh, in decoctions, in juices, in extracts, as latex, in infusions, and as resin.

2.3 Use of Plants as Food

Among the basic requirements of humans, food is the most important one. The population of any part of the world is dependent on the food production capacity of that region. *Triticum aestivum* (wheat) and *Zea mays* (corn), which are the most important crops of Pakistan, are also cultivated in the Lesser Himalayas area. The local inhabitants use the fruits of various plant species, including *Berberis lycium*, *Carissa opaca*, *Debregeasia saeneb*, *Diospyros lotus*, *Ficus* spp., *Myrsine africana*, *Pistacia chinensis*, *Punica granatum*, *Rhus chinensis*, *Rubus ellipticus*, *Segeratia brandrethiana*, *Solanum* spp., *Vitis lanata*, *Zanthoxylum armatum*, and *Zizyphus* spp. in both fresh and dried forms. Most of these fruits are also sold in the local market. Presently these species are under pressure due to population stress and deforestation.

Among vegetables, *Alysum desertorum*, *Amaranthus viridis*, *Bauhinia variegata*, *Brassica campestris*, *B. rapa*, *Chenopodium album*, *Cichorium intybus*, *Ficus* spp., *Laminum amplexicaula*, *Lathyrus aphaca*, *Medicago polymorpha*, *Nasturtium officinale*, *Plantago lanceolata*, *Raphanus sativus*, *Rumex chalpensis*, *R. hastatus*, *Solanum* spp., *Taraxacum officinale*, and *Torilis leptophyla* are commonly utilized species. *Allium cepa*, *A. sativum*, *Coriandrum sativum*, *Capsicum annuum*, *Mentha* spp., *Oxalis corniculata*, and *Pimpinella diversifolia* are used as condiments and spices. The season for collection varies. *Allium cepa*, *A. sativum*, *Brassica campestris*, *B. rapa*, *Chenopodium album*, *Cichorium intybus*, *Ficus variegate*, *Nasturtium officinale*, *Raphanus sativus*, and *Taraxacum officinale*, are collected in early spring, while *Amaranthus viridis*, *Capsicum annuum*, *Medicago polymorpha*, *Mentha* spp., *Oxalis corniculta*, *Pimpinella diversifolia*, *Plantago lanceolata.*, and *Rumex* spp. are collected in early winter. *Allium cepa*, *A. sativum*, *Amaranthus viridis*, *Brassica campestris*, *Coriandrum sativum*, *Ficus variegata*, *Mentha longifolia*, *Oxalis corniculata*, *Pimpinella diversifolia*, *Plantago major*, *Rumex* spp., *Solanum* spp., and *Taraxacum officinale* are also for medicinal purposes, used, for example, to treat asthma, body swelling, body weakness, cholera, cold fever, cough, diarrhea, dysentery, earache, vision weakness, gas trouble, gleets, indigestion, internal pain, internal worms, intestinal and liver inflammation, leucorrhoea, loose stools, menstrual disorders, mouth and gum problems, skin infections, and stomach disorders, to kill germs, and to stop bleeding. The above-mentioned species are not specifically cultivated for diseases. Some species such as *Allium sativum*, *A. cepa*, *Coriandrum sativum*, *Mentha* spp., *Pimpinella diversifolia*, and *Zanthoxylum armatum* are dried in shade for 7–8 days and then stored. These species would be beneficial for the local people if they were cultivated on a large scale, as they provide food and treatment for diseases and could become a good source of income (Figs. 2.1 and 2.2).

Fig. 2.1 A delicious wild fruit

Fig. 2.2 Wild *Pyrus* verity

2.4 Fuelwood Species

Wood is the oldest fuel known to man. Since time immemorial it has been meeting energy needs for domestic activities such as cooking and heating. Until the middle of the nineteenth century, wood was the sole or principal source of domestic and industrial energy worldwide. However, the use of wood as fuel has been steadily replaced by cheaper, more efficient and more convenient sources of energy such as fossil fuels and electricity in developed countries. In developing countries, the process of replacing fuelwood is still in its initial stages, and wood continues to be the dominant fuel for domestic cooking and heating. According to FAO estimates, about 80% of wood removed all over the world is used as fuel in developing countries, and a large majority of rural people and urban poor depend upon it for providing domestic energy. Pakistan has a very small forest resource, as forests cover only about 4.8% of its total land area; only about half of these forests are productive, where timber and fuelwood can be harvested on a sustained basis. Although the foresters' community was advocating for the development and extension of forestry in the country from the time Pakistan gained independence in 1947 to meet the growing needs of fuelwood and timber, nothing substantial was done in this regard until the late 1970s. Initial planning for the establishment of energy plantations on farming lands through farm/social forestry programs was done in the early 1980s. A number of projects were launched by the federal and provincial forest departments in the mid-1980s to promote tree growth on private lands to meet the public's needs for fuelwood [11].

In the Lesser Himalayas, *Acacia modesta, A. catechu, Aesculus indica, Bauhinia variegata, Berberis lycium, Broussonetia papyrifera, Carissa opaca, Celtis caueasica, Cotinus coggyria, Dalbergia sissoo, Debergeasia saeneb, Diospyros lotus, Dodonaea viscosa, Ficus variegata, F. auriculata, Grewia optiva, Juglans regia, Justicia adhatoda, Mallotus phillipensis, Melia azedarach, Morus* spp., *Myrsine africana, Olea ferruginea, Pinus roxburghii, Pistacia chinensis, Populus alba, Prunus* spp., *Punica granatum, Pyrus* spp., *Quercus leuctrichophora, Q. incana, Salix tetrasperma, Segertia brandrethina, Woodfordia fruticosa*, and *Zizyphus jujuba*, are used as fuelwood. Some herbs, such as *Cannabis sativa* and *Zea mays*, are also used for ignition when dry.

Fuelwood is collected by men, children, and, very rarely, women. About 90% of the people depend on plant species for fuelwood (Fig. 2.3), and 5% of residents use kerosene oil and gas cylinders. Residents depend on forest as well as cultivated trees for their fuelwood supply. Today people from the plain areas move to the upper mountains to collect wood, whereas 25–30 years ago, fuelwood was available at their doorstep. This shows the increased deforestation that has occurred in the area during the last 30 years. The main factors responsible for the deforestation include increasing population, fire, and excessive cutting of trees for construction. About 5% of local families make their living selling wood. The wood is collected from far-flung areas (4–5 km from the source of consumption) of the Reserved and Guzara areas in dry (dead) and wet (living) forms. The instruments used for cutting wood are saws (*aree*), axes (*kulahri*), and diggers (*kuddal*). Donkeys and camels are used to transport the wood. Some men even carry the fuelwood themselves. They gather the wood and tie the load with the help of elastic branches of *Mallotus phillipensis, Cotinus coggygria, Myrsine africana, Morus nigra*, and *Dodonaea viscosa.* The method is called *sub*. The wood is used by the houses and small hotels in the area. Fuelwood is also stored; this storage is done in the rainy season (July–September).

Some species, including *Dodonaea viscosa, Mallotus philippensis, Myrsine africana, Olea ferruginea, Pinus roxburghii, Punica granatum*, and *Quercus leuctrichophora* are under high pressure due to increased human population. It has been observed that more wood products are found at places that have been declared

Fig. 2.3 Fuel wood

sacred (shrines, graveyards, and rakhs). Locals pay great homage particularly to the shrines. These intentions of the people can be suggested to use for the protection of reserved areas if the areas near the shrines may be dedicated to the name of the respective saint or shrines by preaching through local religious people. As local inhabitants give more respects to such place and don't exploit vegetation so it is suggested that we can protect vegetation of a place, if we dedicate that place to the name of respective saint or shrines.

2.5 Fodder Species

Forest grazing, a conventional resource, follows a centuries-old use of the forestland in Pakistan. Almost all types and legal categories of forests are burdened with unspecified grazing rights and privileges. The grazing pressure has been increasing with the increase in human and livestock populations. Consequently, uncontrolled heavy grazing is causing great damage to soil and vegetation due to compaction and trampling. This creates gaps in the forests and retrogression in certain localities. Summer grazing by both local and nomadic livestock is very common in the moist temperate forest ranges in the northern mountainous tract. These forests are mostly located between 200 and 300 m above sea level, where, due to favorable moisture and temperature conditions, luxurious ground vegetation, and perennial and annual grasses, heavy uncontrolled grazing causes considerable damage to both the forest and range vegetation [12].

Fig. 2.4 Fodder for live stock

Fodder is the basic demand of cattle. Cattle's fodder requirement is fulfilled in the area because it is rich in fodder species. Local people use trees, shrubs, and herbs as fodder for their livestock. Primarily the members of the family Poaceae are used as fodder in both fresh and dry forms throughout the year. However, *Brassica campestris* (Sarsoon), *Berberis lycium* (Sumbal), *Celtis caucasica* (Batkair), *Ficus* spp. (Phagwara), *Grewia optiva* (Dhaman), *Melia azedarach* (Drek), *Morus* spp., and *Punica granatum* (Drauna) are used as fodder in their respective seasons. Grazing is the usual practice for cows, goats, and buffalo. These domestic animals fulfill the dairy requirements of the local people as well as improve the local micro economy. Agricultural areas and "Rakhan" consist of those areas of Guzara forests that have abundant grasses, are harvested in September for the winter, and are regulated as the property of households not cut before September and protected from grazing. Rakhan areas form patches of Guzara forests that are the only source of fodder in winter.

Grasses are the most important fodder of the area and are found abundantly. Major grass species that are grazed by the animals and stored for winter are *Alopecurus myosuroides*, *Aristida cynantha*, *Avena sativa*, *Cynodon dactylon*, *Dichanthium annulatum*, *Heteropogon contortus*, *Phlaris minor*, and *Sorghum halepense*. Grasses are stored from August to October after the monsoon rainy season (Fig. 2.4). The stored grasses are used from November to January. From mid-January, when the stored fodder diminishes, the leaves of trees and shrubs, mainly *Grewia optiva* (Dhaman), *Celtis australis* (Batkair), *Quercus* spp., *Broussonetia papyrifera* (Gangli toot), *Ficus variegate* (Phagwari), *Myrsine africana* (Khukan), *Segeretia brandrethiana* (Ghangir), and *Olea ferruginea* (Kahu), are used, which play a vital role in maintaining the fodder supply during the off-season (November to February).

The grasses are cut with a sickle (*dranti*) and spread over the land for drying in handfuls called *datha*. After the *dathas* are dry, people make one *gadi* of 12 *dathas* and put it back on the land for further drying for 3–4 days. Then they make *gada*, which is a rectangular stack by putting six to eight *gadis*, one upon the other. Then, by stacking *gadas* one upon the other, they make a *ghara* (the complete stock), which has a rectangular shape. *Sorghum halepense* grass (Baru) is placed on it, so that rainwater cannot enter. Ropes (*sub*) of *Aristida funiculata* (Bhari) and *Dichanthium fovealatum* (Palwa) are used to tie the *ghara*. In the case of maize, people cut the maize and spread on land for 3–4 days. When it is dried, they tie the stocks in armfuls called *poola* and spread them on the land. With the help of wood, they make *ghori*, which consists of two Y-shaped pieces of wood created on the land with a straight piece of wood placed between them. They place the *poola* along this on both sides, along the length of the straight piece of wood. When the *poola* are dried, they make *ghara*. The fodder species are found in the area near houses; sometimes people have to go 2–3 km away from their houses. Mostly, male collect the fodder, but females also participate. Women usually conduct one trip per day during the collection period. They also conduct two to three trips in exceptional cases (marriage and festivals). According to area inhabitants, the density of grasses and other fodder species is decreasing due to the loss of soil fertility, increase in population, fire, overgrazing and browsing, and increased use of plants as fuelwood. Local people suggest that grazing is necessary for growth of the grasses as the feces of animals provides very good manure, preventing the soil from becoming nutrient-deficient.

Local people also use trees and shrubs as fodder for their livestock, including *Carissa opaca*, *Berberis lyceum*, *Broussonetia papyrifera*, *Diospyros lotus*, *Ficus variegata*, *Quercus leuctrichophora*, *Acacia* spp., *Morus* spp., *Melia azedarach*, *Olea ferruginea*, *Zizyphus jujuba*, *Myrsine africana*, *Pyrus* spp., *Prunus* spp., *Grewia optiva*, *Segeretia brandrethiana*, and *Celtis australis*. They cut branches with leaves with the help of sickles. The basic unit of collection is called a *phant* or *dali*. Thirty to 40 *phants* or *dalis* are tied together with the help of elastic branches of *Morus* spp., *Dodonaea viscosa*, *Myrsine africana*, *Grewia optiva*, *Vitex negundo*, *Olea ferruginea*, and *Cotinus coggyria* called *sub*. Then they use *sub* to tie the entire *phants* or *dalis* together to make a *gada*. The weight of this *gada* is about 25–35 kg. Among households interviewed, it was found that most of them could meet their winter fodder requirement. But those who could not meet it buy grass, maize, and wheat stalks from local villages. Livestock, especially buffalo, cows, and goats, totally depend upon the stored fodder and fodder species found during winter.

Grazing animals provide a very good natural material for the soil that ensures the regeneration of fodder species next year. However, locals may be told to avoid periodic grazing of specific areas, to give that area enough time for recovery. This can be achieved through rotational grazing; based on community self-management, it encourages keeping livestock of improved breeds and helps in the formation of livestock associations. *Olea ferruginea*, *Myrsine africana*, *Accacia* spp., *Quercus leuctrichophora*, *Morus* spp., *Pyrus* spp., and *Grewia optiva* emerge as the most sustainable.

These species play a vital role in maintaining the fodder supply during the off-season of November to February. The wood, branches, bark, and fruit of these species are used as food, fuel, agricultural tools, rope, and thatching, among other uses.

2.6 Agricultural Implements

A proper tool handle is one of the basic requirements for the safety and high productivity of forest workers. Several forest tools have wooden and metallic parts. Substantial amount of work has been done throughout the world to design tool handles, which fulfill the agronomical and physical requirements of the job. In the case of wooden handles, the choice of the species depends upon its strength and other desirable characteristics. Beside strength and elasticity, other properties such as smoothness and the type of splintering that takes place during the failure of a handle are also important [14]. Inhabitants of the area use different plants species in making agricultural implements, ploughs, tools handles, sticks, sickle, dagger, hoe, axe and knife handles. They are made from locally available hard and soft wood. *Quercus leuctrichophora, Dalbergia sissoo, Morus* spp. *Diospyrus lotus, Acacia* spp. *Juglans regia, Melia azedarach* and *Olea ferruginea* are among the commonly utilized species. The most preferred wood is that of *Quercus leuctrichophora, Olea ferruginea, Morus alba, M. nigra, Melia azedarach, Psiticia integrrima, Salix tetrasperma, Populus alba, Diospyros lotus, Dalbergia sissoo* and *Acacia* spp. (Fig. 2.5).

Fig. 2.5 Wood for construction purpose

2.7 Fencing and Hedges

Spiny and bushy species are used for fencing and hedges. These species are cultivated on the margins of fields and form a permanent fencing or branches of these plants are fixed in mud on the margin and form temporary fencing. *Carissa opaca*, *Berberis lycium*, *Acacia* spp., *Punica granatum*, *Zizyphus* spp., *Rubus* spp., and *Zanthoxylum armatum* are among the commonly utilized species for fencing and hedges.

2.8 Construction Material

In overall botanical importance to human existence, only food plants rank above wood and wood products. In early human history wood had been of greater importance than the food plants, as a fuel and for weapons and tools. There are over 4500 products that come wholly or in part from the wood of forest trees. Wood is used for housing, furniture, fuel, paper, charcoal and distillation by products, and synthetic materials such as rayon, cellophane and acetate plastics. One of the most obvious uses for trees is the production of building and furniture. Many millions of board feet of certain softwoods are used each year for home construction. Because of the grains, colors and durability of hardwoods, they are most often used in furniture making. Now a day is introduced in house construction like concrete roofs, iron doors and windows but plants still play very important in the construction of homes. The most preferred species used for the said purpose are *Acacia* spp, *Pinus roxburghii, Myrsine africana, Dodonaea viscosa, Dalbergia sissoo, Olea ferruginea, Morus alba, M. nigra, Celtis caucasica, Psiticia integrrima, Segeretia brandrethina and Quercus* spp. Leaves, branches and poles of these species are used, as most of the local houses have mud roofs. *Pinus* spp., *Morus spp. Quercus spp., Melia azadarach* and *Olea ferruginea* wood is also sold in the local markets as timber.

2.9 Miscellaneous Uses

Cotinus coggyria, Myrsine africana, Juglans regia, Pinus roxburghii, Acacia spp., *Triticum aestivum, Grewia optiva, Celtis australis, Dodonaea viscosa, Phoenix sylvestris, Grewia optiva, Salix tetrasperma* and *Berberis lyceum* are used miscellaneously. Women and children make ropes, bags, baskets, mats, ornamental goods and handicrafts from these plants which are used in daily life. The use of non-timber forest products like wild fruit and flowers for food, handicraft making, mats, dry decoration pieces from leaves, and rope making from bark should be encouraged and properly managed by the local social organizations for better use of resources for benefit of the local people and protection of ethnobotanical culture.

2.10 Grafting

Grafting is already being done quite successfully by local people in late winter (February–March) on *Morus* spp., *Ficus* spp., *Pyrus* spp., and *Prunus* spp. to get higher fruit production to improve their micro economy. There is also the possibility of grafting *Olea europea* (olive) on the native *Olea ferruginea* (wild olive) to make it a more profitable and sustainable species for the local people. In this regard, successful experimental trials were carried out at National Agriculture Research Centre (NARC) in Islamabad. But the results have yet to be applied in the Lesser Himalaya areas. Grafting different strains of *Olea europea* from different countries should be done to make it successful.

2.11 Beekeeping

Beekeeping is currently becoming a most beneficial industry (Fig. 2.6). The flora of the Lesser Himalayas has great potential for honey beekeeping. A number of wild species are normally visited by the bees, including *Acacia modesta*, *Bauhinia variegata*, *Brassica campestris*, *Carissa opaca*, *Dalbergia sissoo*, *Justicia adhatoda*, *Punica granatum*, *Pyrus* spp., *Prunus* spp., *Zea mays*, Beekeeping has been identified as a small-scale, nonland-based, off-farm activity that can facilitate the use of land resources without degrading them. Encouraging this industry in the area will stress to the local community the importance of keeping the flora alive as

Fig. 2.6 Apiculturing

well as being cautious about starting fires in the local forest, as fumigation badly disturbs beekeeping. Older people and women of the area can undertake this activity as a domestic industry. The association of wild animals with the is a natural phenomenon; for example, according to the older generation, there was a thriving population of barking deer when the vegetation of *Dodonaea viscosa, Myrsine africana, Mallatous Phillipensis*, and *Olea ferruginea* was dense, but due to habitat degradation, the barking deer population has also diminished. Several other animal species have become extinct, but they should be reintroduced to the area after rehabilitation.

2.12 Marketing of Medicinal Plants

The marketing of crude herbal drugs needs special attention due to their widespread use by traditional practitioners of the Greco-Arab system of medicine. Approximately 5,000 poor families residing in the remote hilly areas are engaged in the collection of medicinal plants during the summer months in the northern regions. Medicinal plants are transported to other markets by the seasonal traders, from where these commodities ultimately find their way to other parts of the country for consumption and export [15]. Medicinal plants are used not only by local practitioners as household remedy but also provide raw material for the pharmaceutical industries of the country. The business is in the hands of a few large trading houses in the areas that neither are organized nor work along scientific lines for the collection, drying, cleaning, washing, storage, and standardization of medicinal plants. The medicinal plants from the study areas are collected by local inhabitants, drug dealers, village grocers, and local practitioners through traditional knowledge and having no scientific background and approach (Fig. 2.7). They collect every possible available part of the plant. Most of the collectors are also ignorant of, or have insufficient knowledge about, the proper time of collection. Medicinal plants are either dried for further use or sold directly to the local grocers in fresh form, where the grocer does the drying him- or herself. Women generally do drying at home by spreading the plants on the floor, plastic sheets, cloth pieces, mats, and so forth, both in sunlight and in shade. The drying takes about 4.5 days. The quality of the drying process is generally very poor, as dust and foreign materials get mixed in with the plants. The fresh plant materials on the market are dried by being spread in sunlight for 4–6 days and then graded (pure, mixed), packed, and stored in bags ranging in volume from a few kilograms to mounds (1 mound = 40 Kg), depending upon the mass and availability of the drug. There is no storage process at the collector's level, because they try to sell them as soon as possible. Village grocers have to store small quantities for a short while until they are able to sell them to wholesalers of the local markets in Rawalpindi, Abbottabad, Murree, and Haripur. Like drying, the storage is not done in hygienic conditions, and the crude drug often gets infected with insects and fungi. This results in the deterioration

Fig. 2.7 A local herbal dealer

of the dried crude drugs and ultimately causes financial loss to the traders. In order to maintain quality, storage facilities need a definite improvement.

Rawalpindi, Abbottabad, Murree, and Haripur are the important markets for crude drugs. They are easily available at cheap prices. The wholesalers sell the drugs to bigger markets such as Lahore and Karachi, from where it is exported. Among the commercially exploited drugs *Viola canescens, Olea ferruginea, Pistacia chinensis, Zanthoxylum armatum, Acacia modesta, Brassica campestris, Mallotus phillipensis, Berberis lycium* and *Bergenia ciliata* fetch high price, and the rates of *Viola canescens, Berberis lycium, Acacia modesta, Bergenia ciliata*, are going up due to the shortage of their availability. There is a great difference between the purchase and sale prices. Grocers, shopkeepers, and wholesalers explain that collectors bring fresh drug, which, upon drying, becomes lighter in weight; they also include the labor cost they pay. They also claim adulteration of other plant parts and extra material by collectors, which they have to remove, and the drug packed becomes lighter in weight, which in turn increases its price. The people in the study areas have not yet tried to cultivate or extensively grow these medicinal plants and later on sell them in markets or send them to pharmaceutical industries. Presently, there is no well-organized system for the cultivation of medicinal plants on farms. The cultivation of medicinal plants is generally not practiced in comparison to agricultural crops in the study areas. This might be due to local people's unawareness regarding the medicinal uses, cultivation, seed collection, sowing, harvesting, collection, storage, and marketing value of medicinal plants. There are a lack of information regarding the marketing value of these medicinal plants and also a carelessness on the government's part. The other reason for not cultivating these medicinal plants is the lack of land. Most of the people in the study areas possess as little as 4 acres of land. Medicinal plants can provide better income to the local people than the traditional crops of the areas, if the market system for the medicinal plants is improved and cultivation of medicinal plants is done on scientific lines [13].

2.13 Anthropogenic Pressure

Due to overuse, medicinal plants are vulnerable and are considered rather threatened since their population is thinly scattered and cannot be commercially utilized on a large scale. The entire plant of *Ajuga bracteosa*, *Berberis lyceum*, *Bergenia ciliata*, *Viola canescens*, and *Zanthoxylum armatum* is used. Due to demand, the existence of these species will be threatened in the future. The only way to protect these plant species is to make the local communities aware of well-managed propagation and regeneration techniques. Although the fruit products of *Punica granatum*, *Prunus* spp. and *Pyrus* spp. are being sold on the local market, these species face pressures such as being cut down for fuel, furniture, and fodder purposes. Yet the species' density in the areas is satisfactory and can be sustained if other pressures like being cut down for fuel and furniture are removed. The case is similar with *Carissa opaca*, *Myrsine africana*, and *Mallotus phillipensis*. *Berberis lycium*, *Bergenia ciliate*, *Justicia adhatoda*, *Pistacia chinensis*, *Quercus leuctrichophora*, *Punica granatum*, and *Viola canescens*, are vulnerable due to their part(s) used, growth rate, quantity of consumption, and pressures like grazing, erosion, and fuelwood collection. They in particular need to be conserved by domestication and regeneration techniques.

2.13.1 Fire

Fire is the most important factor that damages the vegetation over large areas. Fire not only destroys plant regeneration but also replaces trees with inferior scrub-type broad-leaved species such as *Berberis lycium*, *Mallotus phillipensis*, *Myrsine africana*, and *Rhus cotinus*. In most of cases, the outbreak of fire is deliberate and is set by villagers in order to get the forest floor clear of pine needles to facilitate the movement of their cattle and to get a good growth of grasses, which come up abundantly after fire. People who have their own lands in forest areas or who live in forests are involved in fires. The fire normally continues for 5–6 days. Although *Pinus* spp. is a fire-hardy tree, great damage still occurs to the young crop as well as to mature trees by the presence of inflamed resin blazes. Repeated ground fires also cause soil to dry up and create drought conditions. In some cases, the trees are not apparently burned, but are subsequently killed by fire after-effects, such as desiccation and drought. The greatest damage is done to the young generation of *Pinus* spp., *Dodonaea viscosa*, *Diospyros lotus*, *Myrsine africana*, and so forth.

Fig. 2.8 Himalayan nomads

2.13.2 Browsing

Browsing by goats and sheep is another problem. The absence of regeneration in most of the areas, especially in brushwood forests, is solely due to this unwanted practice. The population of the area has free and unrestricted rights for the free grazing and browsing of domestic cattle. Large flocks of goats and sheep from the Kaghan Valley come down to these forests during winter in addition to already existing goats and sheep in the villages. The privately owned areas for browsing are limited, and more often illicit browsing is done in the forest areas (Fig. 2.8).

2.13.3 Grazing

Local domestic animals graze in the forest areas without any check, which results in the destruction of vegetation and also encourages the establishment of inferior species. Heavy grazing is harmful because it accelerates the washing down of rich top soil and thus renders conditions more unfavorable for natural regeneration (Fig. 2.9).

Fig. 2.9 Grazing pressure

2.13.4 Grass Cutting

Grass cutting in the pine zone is beneficial if it is carried out carefully because it reduces fire hazards in addition to being used as fodder for the local cattle. But the grass cutters never take care of the young seedlings being cut down. The damage during grass cutting is much less in brushwood areas.

2.13.5 Lopping

Lopping of the *Quercus* spp., *Acacia modesta*, and *Olea ferruginea* species up to two thirds of the tree's height is allowed under Guzara rules. But it is observed that local people do not follow these instructions, and they not only lop up to the top, but the poles of *Olea ferruginea*, *Quercus leuctrichophora*, *Diospyros lotus*, and *Pinus* spp. have been seen to be cut in the middle and at the top, resulting in the trees' death.

2.13.6 Torchwood

The local people have the pernicious habit of hacking out the resinous or torchwood *Delhi* from mature or nearly mature pine trees. During the search for proper wood, several trees are injured (Fig. 2.10) before suitable ones are found for subsequent extraction. These forest areas, which are generally in the neighborhood of inhabitants, suffer a great deal from this damage.

Fig. 2.10 Torch wood extraction

2.13.7 *Climatic Factors*

Among climatic factors, the important ones are snowfall, wind, drought, and frost. Snowfall in the area is not heavy and does not damage vegetation. The areas are also not subjected to violent windstorms, but occasional damage from high winds may occur to solitary pine trees, which were left naked at the base for torchwood or severely damaged by fire. Ordinary frosts are not severe in winter, but *Dodonaea viscosa*, which is frost tender, does suffer to some extent from frost.

2.13.8 *Wild Animals and Insects*

The damage to vegetation caused by wild animals and insects, like beetles, porcupines, and wild pig, is not of significance in the area.

2.13.9 *Agriculture*

The people of the Lesser Himalayas are generally agriculturists. Due to the limited and small, uneconomical land holdings, they usually seek employment locally as well as in Islamabad, Abbottabad, Murree, Rawalpindi, Haripur, and Wah. In recent years, the growing trend has been to seek employment in the Gulf States. Since the people of the area are agriculturists, in order to obtain more food grains for their increasing family size, they put more land under cultivation. Local people classify agricultural land into two basic units: (1) *kalsi* (contains 8 *marlas* (1 marla = 272 sq.ft.)) and (2) *Doga* (contains more than 8 *marlas*). For the cultivation of crops the plowing of land for the first time is called "*khili patna*". People plow the land again the same day if the land is a small area; larger areas are plowed the next day. Then they plow the land a third time before spreading the seeds on the land. They call this plowing *rai karna*, which means "sowing seeds." If the land is one *kanal* (1 kanal = 20 marlas (5440 sq. ft)), then plow the land from one side to the other is called *ang*. Three kilograms of maize seeds and 6–8 kg of wheat seeds are used for one *ang*. About 8–10 kg of maize and 10–15 kg of wheat seeds are used on one *kanal*. The plowing of land in hilly areas is mostly done with oxen in pairs. People who do not have oxen hire them for a day or more according to their needs. The cost of hiring oxen per day is 600–700 rupees (Rs). In the plain areas where the land is flatter, land is plowed with a tractor leased at 700 Rs/h. There is a lack of good, cultivable land. The people who live in the hilly areas usually have much land, but the productivity of this land is very low.

2.14 Recommendations

It is evident from the present investigation that the Lesser Himalayas is a rich area particularly with reference to medicinal plants. Its vegetation is valuable due to its natural resources. The natural resources must be looked after and managed. In order to conserve these resources, the local people must become actively involved in the evaluation, planning, implementation, and monitoring processes, as they are the best judges of the area.

The following recommendations are being proposed to conserve the plant species and to reap the greatest benefits from the available resources:

- The literacy rate must be increased; this will change people's minds about current practices.
- Local organizations may involve local people as leaders of activities geared toward environmental conservation awareness. Local schoolteachers and religious leaders must be considered for such positions.
- In the present situation, the authorities should give attention to the sustainable use of resources.
- The locals should be educated about the importance of medicinal plants to their socioeconomic conditions and to the ecosystem.
- Small domestic industries such as beekeeping, gardening, handicrafts, and so on must be encouraged in the area through social organizations within the local communities.
- Forest rules must be overhauled by taking villagers into confidence, as misuse of the present rules has been reported.

- Botanical gardens of local medicinal plants should be made in the area where both the folk plants and their lore can be displayed for visitors.
- Research and postgraduate educational institutions should be involved to point out regeneration and propagation techniques for vulnerable medicinal species.
- The collection of medicinal plants carried out by locals may be streamlined in a way that provides ample regeneration time to the plant, keeping its optimum period of growth in mind. The area once used for collection may be declared a protected area, with extraction prohibited for a few years.
- The reforestation of fuelwood and fodder species must be encouraged, and alternate sources like gas cylinders, energy-efficient cookstoves, and tandoor ovens should be made available to local people.
- The reserved forests around religious shrines may be dedicated to the name of the respective shrine or saint.
- Improved livestock husbandry could significantly reduce grazing pressure in the area. Rotational grazing should be encouraged against periodic overgrazing among the local communities.
- Awareness and incentives for planting locally useful trees may be launched by schoolteachers, leaders of local mosques, and village elders.
- Further research should be done on the breeding biology, extent of natural range, threats, and population dynamic of endangered, rare, and endemic species of wild plants and animals.
- A regional conservation committee should be created for the area.
- Researchers and medicinal plant experts should visit the area during March to August for plant studies.
- In order to avoid further loss of endangered, endemic, and rare species, conservation methods should be practiced as part of a long-term conservation program.
- Reforestation trends have been lacking among the local communities. Along with regeneration activities, an alternate source should be provided to reduce pressure on fuelwood. For instance, for lower-income people, energy-efficient cookstoves, gas cylinders, and tandoors can be provided, leading to a 25–40% fuel saving.

References

1. Pei SJ (1992) Mountain culture and forest resource management of Himalaya. In: Tewari DW (ed) Himalayan ecosystem. International Book Distributor, Dehra Dun, pp 125–128
2. Hamilton A (2002) Curriculum development in applied ethnobotany. Proceedings of the workshop on curriculum development in applied ethnobotany, Nathiagali, pp 91–95
3. Olsen O, Svanberg I (1998) Tanning with tormentil (*Potentilla erecta*): ecological and ethnobotanical aspects, Studia ethnobiologica 1. Swedish Biodiversity Center, Uppsala
4. Cotton CM (1996) Ethnobotany: principles and applications. Wiley, Chichester
5. Clement D (1998) The historical foundations of ethnobiology. J Ethnobiol 18:161–187
6. Balick MJ, Cox PA (1996) Plants, people, and culture: the science of ethnobotany. Scientific American Library, New York

7. Gilani AH, Mehmood MH, Janbaz KH, Khan AU, Saeed SA (2008) Ethnopharmacological studies on antispasmodic and antiplatelet activities of *Ficus carica*. J Ethnopharmacol 119:1–5
8. Turner NJ (1995) Ethnobotany today in northwestern North America. In: Schultes RE, von Reis S (eds) Ethnobotany: evolution of a discipline. Dioscorides Press, San Francisco, pp 264–283
9. Pei SJ (2002) Bio-cultural diversity and development of western China. Journal of Graduate School of the Chinese Academy of Sciences 19(2):107–115
10. Gu Q, Wang RR, Zhang XM, Wang YH, Zheng YT, Zhou J, Chen JJ (2007) A new benzofuranone and anti-HIV constituents from the stems of *Rhus chinensis*. Planta Med 73(3):279–282
11. Siddiqui KM (1990) Wood as a source of energy in Pakistan (current situation and future prospects). Pakis. J Forest 40(4):261–266
12. Rafique SM (1994) Effect of grazing management and fertilizer application on vegetation and soil properties of a moist temperate forest range in Siran Valley (Mansehra), NWFP (Pakistan). Pakis. J Forest 44(4):20–29
13. Latif A, Shinwari ZK (2005) Sustainable market development for non timber forest products in Pakistan. Ethnobotanical leaflet
14. Northington DK, Goodin JR (1984) The botanical world. Wm. C. Brown Publishers, Oxford, pp 227–595
15. Zadi SH (1989) Position paper on marketing of crude herbal drugs and their constraints. Proceedings of workshop on wild plant resources of Northern Pakistan. Pakistan Forest Institute, Peshawar, pp 33–35

Chapter 3
Medicinal Plants Inventory

Abstract This chapter includes complete inventory of 100 medicinal plant species belonging to 3 genera, 3 species and 3 families of pteridiophytes, 2 genera, 3 species and 1 family of gymnosperms, 11 genera 12 species and 7 families of angiosperms (monocots), 71 genera, 82 species and 52 families of angiosperms (dicots). Complete profile of individual medicinal plant i.e. botanical name, family, vernacular name, flowering period, status, part/parts used, habit, habitat, distribution, description, medicinal uses (collection, recipes and diseases cured), ethnobotanical uses and phytochemicals is provided.

Keywords Angiosperms (dicots) • Angiosperms (monocots) • Description • Distribution • Ethnobotanical uses • Gymnosperms • Medicinal uses • Phytochemicals • Plant inventory • Pteridophyes

3.1 Section A Pteridophytes

3.1.1 Adiantum capillus-veneris Linn.

Family Name:	Adiantaceae
Local Name/ English Name:	Hansraj, Sraj, Pershan, Kakapi/Maidenhair fern
Spore Period:	May–August
Part Used:	Leaves
Conservation Status:	Common
Habit/Habitat:	A wild herb, found in cold, moist, and shady places along watercourses in moist clay.

(continued)

DOI 10.1007/978-1-4614-1575-6_3,

Fig. 3.1 *Adiantum capillus-veneris* Linn

Family Name:	Adiantaceae
Distribution:	***Pakistan:*** Sind, Peshawar, Jhelum, Khari Murat Hill, Attock, Balouchistan, Salt Range, Hazara, Mansera, Batrasi Pass, Kashmir, Kaghan, Naran, Murree, Abbottabat, and Balakot. ***World*:** British Columbia, southern U.S., Mexico, Central America, Venezuela, Peru, Chile, Greater and Lesser Antilles, Barbados, Trinidad, Europe, Asia, Africa, Comoro and Mascarene Islands, Hawaii, and Australia.
Description:	The main plant body sporophyte, differentiated into roots, stem, and leaves. Roots black, adventitious, arise from the underside of creeping stem. Stem underground rhizome, long and creeping, some time erect, covered with scales. Leaves compound, bipinnate, alternate or spirally arranged on the stem; young leaves circinate. Each leaf has a long, shiny, black, and brittle petiole. Leaflets alternate or opposite, sessile. Rachis black, hard, and 3–40 cm in length. Sporangia with spores are present on underside of the leaves (Fig. 3.1).

(continued)

Family Name:	Adiantaceae
Medicinal Uses:	
Collection:	250 g of fresh plant (leaves) is collected in any season by men and women 20–40 years old. Then the leaves are cleaned and washed in water 2–3 times, They are used in both fresh and dried forms; they are dried in shade for 2–3 days and stored in cotton sacks or baskets for further use.
Recipes:	200 g of fresh plant material is boiled daily in 500 mL of water for 15–20 min. Half a cup (50–60 g) of sugar is also added. When 3 cups (750 mL) of water are left, the water is filtered with a cloth and given to the patient. **For children,** ½ cup (125 mL) of decoction (at one time) is given twice daily (morning–evening) for 8–10 days. **For adults,** 1 cup (250 mL) of decoction (at one time) is given 2–3 times per day for 10–15 days.
Diseases Cured:	Cough, asthma, fever, jaundice, cold, chest pain, and measles.
Phytochemicals:	Kaempferol, quercetol, luteol, adiantone, esters, and oxohakonanol [1].

3.1.2 *Equisetum debile Roxb.*

Family Name:	Equisetaceae
Local Name/ English Name:	Satgandi booti/Chinese herb, horsetail
Spore Period:	August–September
Part Used:	Whole plant
Conservation Status:	Common
Habit/Habitat:	A wild perennial herb, mostly found in the banks of streams, nallahs, in sandy and swampy soil.
Distribution:	***Pakistan:*** Chitral, Dras, Swat, Hazara, Murree, and Kashmir. ***World:*** India, Sri Lanka, Himalaya, Nepal, Burma, Taiwan, Hainan, New Guinea, Vanuatu, New Caledonia, Fiji (Viti Levu), Philippines, Sri Lanka, Laos, Cambodia, Vietnam, Indonesia, peninsular Malaysia (Gunung Ulu Kali and near Raub, Pahan), Borneo (Mt. Kinabalu, etc.), Thailand, Sumatra, Java, Lesser Sunda Islands, Sulawesi, and Moluccas.

(continued)

Fig. 3.2 *Equisetum debile* Roxb

Family Name:	Equisetaceae
Description:	Main plant body differentiated into root, stem, and leaves. Roots thin, fibrous, much branched, underground arising in whorls from each node of underground rhizome. Stem underground dark brown rhizome, creeping, much branched, and differentiated into nodes and internodes. Arial shoots green, angular, stiff, and rough arise from nodes and grow upward and bear nodes and internodes. Leaves small, brown, simple, scale-like, and arise in whorls from each node of rhizome and aerial branches (Fig. 3.2).
Medicinal Uses:	
Collection:	100–150 g of fresh plant material is collected daily by women 25–40 years old, in summer (April–July). Then it is cleaned and washed in water 2–3 times and cut into small 1–2-in. pieces.

(continued)

Family Name:	Equisetaceae
Recipes:	100 g of fresh plant material and 20–25 g sugar are ground together for 5–6 min. Then 1 cup (250 mL) of water is mixed with it. This mixture is filtered with a piece of cloth and given to patients suffering from jaundice, liver, and intestinal inflammation. **For children,** 1 cup (250 mL) of drug (at one time) is given once daily, before breakfast, for 7–8 days. **For adults,** 1 cup (250 mL) of drug (at one time) is given once daily, before breakfast, for 10–15 days.
Diseases Cured:	Jaundice and intestinal inflammation.
Phytochemicals:	Moisture, ash, lipid, protein, carbohydrates, Na, K, Ca, P, Cb, Cu, Ni, Zn, Cr, ascorbic acid, folic acid, niacin [2].

3.1.3 Pteridium aquilinum (L.) Kuhn

Family Name:	Pteridaceae
Local Name/English Name:	Ghanduli booti/Bracken fern
Spore Period:	May–August
Parts Used:	Rhizome, hairs
Status:	Common
Habit/Habitat:	A wild herb, mostly grows in moist and shady places in mountains, forest floors, grooves, exposed grasslands, and in dry open places in clay loam.
Distribution:	***Pakistan:*** Swat, Dir, Chitral, Hazara, Malakand, and Haripur. ***World:*** Temperate and subtropical regions throughout the world, including most of Europe, Asia, and North America, in the Northern Hemisphere, and Australia, New Zealand and northern South America.
Description:	Small wild herb up to 2 ft in length. Roots adventitious; arise from the lower surface of rhizome. Stem horizontal, modified into underground rhizome, thickly coated with pale brown multicelluar hairs. The rhizome differentiated into nodes and internodes. Leaves macrophyllous, compound, 2–12 ft in length, arise on the upper side of rhizome, while young leaves are coiled and covered with brownish hairs; petiolate bears many green leaflets. Brown-color spores bearing sporangia are present on lower surface of leaf (Fig. 3.3).

(continued)

Fig. 3.3 *Pteridium equilinium* (L.)

Family Name:	Pteridaceae
Medicinal Uses:	
Collection:	30–35 g of fresh rhizome is collected in any season by men and women between 20–40 years old. Then hairs are removed with bare hands.
Recipes:	**(a)** 25–30 g of fresh rhizomes is boiled daily in 1 cup (250 mL) of water for 15–20 min; when 1 cup (250 mL) of water remains, it is filtered with a piece of cloth and given to patients suffering from diarrhea, vomiting, and dysentery. **For children,** 1 tsp (5 mL) of decoction (at one time) is given once per day for 2–3 days. **For adults,** ½ cup (125 mL) of decoction (at one time) is given twice daily (morning–evening) for 3–4 days.

(continued)

Family Name:	Pteridaceae
	(b) 25 g of fresh rhizome and 2–3 tsp (20–25 g) sugar are ground together daily for 4–5 min, mixed in 1 cup (250 mL) of water, and filtered with a cloth. This syrup is given to patients suffering from body pain due to swelling, muscular pain, earache, mouth sores, and sore throat, and to purify blood. **For children,** ½ cup (125 mL) of syrup (at one time) is given 2–3 times per day for 3–4 days. **For adults,** 1 cup (250 mL) of syrup (at one time) is given 3 times per day for 4–5 days.
Diseases Cured:	Body pain due to swelling, muscular pain, earache, mouth sores, sore throat, vomiting, diarrhea, dysentery, and to purify blood.
Ethnobotanical Uses:	Hairs are burned on a fire and curd pots are placed on smoke. According to the local women, smoke kills the pot's germs.
Phytochemicals:	Carcinogenic (ptaquiloside), cynogenic glycoside (prunasin), vitamin B1, and thiaminase enzyme, thiamine [3].

3.2 Section B Gymnosperms

3.2.1 Abies pindrow Royle

Family Name:	Pinaceae
Local Name/English Name:	Partal, Plundar, Achar/Silver fir
Flowering Period:	April–September
Status:	Rare
Parts Used:	Stem, bark, wood
Habit/Habitat:	A characteristic tree, grows in subtropical pine forests.
Distribution:	***Pakistan:*** Swat, Hazara, Dir, and Galiyat. ***World:*** Afghanistan, Himalayas eastward to W. Nepal.
Description:	Tree up to 30 m tall or more, with a narrow pyramidal shape. Bark light gray to brown. Leaves spiral, dark green, shiny. Male cones 1–2 cm long, axillary, reddish-green. Female cones solitary or in pairs, violet-purple. Seeds wing twice as long as the seed (Fig. 3.4).

(continued)

Fig. 3.4 *Abies pindrow* Royle

Family Name:	Pinaceae
Medicinal Uses:	
Collection:	200 g of fresh leaves is collected in any season by men 20–40 years old; the leaves are cut into small 1–2-in. pieces.
Recipes:	Dried leaves are cut into small pieces, boiled in water, and given to patients. **For children,** ½ cup (125 mL) of drug (at one time) is given 3–4 times per day for 2–3 days. **For adults,** 1–2 cups of decoction is used twice a day for 8–10 days.
Diseases Cured:	Cough, asthma, chronic bronchitis, other pulmonary afflictions, and catarrh of the bladder.
Ethnobotanical Uses:	The plant is a useful timber tree for building purposes, used in floors and roofs of houses, vehicles, and furniture. Also used for making matches and paper pulp. Along with *Taxus,* it is also used in gunpowder. The wood is also extensively used as fuel. The dried stumps are very oily and used for fuel and light purposes. The trees are tallest in the forests. They are umbrella-like and provide thick shade and shelter for wild animals, goats, sheep, and shepherds, especially during rainy seasons.
Phytochemicals:	Tricosane, eicosane, heneicosane, docosane, tetracosane, nonadecane, octadecane, 1-docosene, 1-octadecene, heptadecane, and 2,6,10,14-tetramethylhexadecane [4].

Fig. 3.5 *Pinus roxburgii* Sargent

3.2.2 Pinus roxburghii Sargent

Family Name:	Pinaceae
Local Name/English Name:	Chir, Nakhtar/Chir pine
Flowering Period:	February–April
Status:	Common
Part Used:	Whole plant
Habit/Habitat:	A wild evergreen tree, found on hills as self-growing plant in loamy soil.
Distribution:	***Pakistan:*** Murree, Hazara, Swat, Dir, and Kashmir. ***World:*** Afghanistan, India, and Bhutan.
Description:	Evergreen tree up to 40 m tall. Stem erect, woody, with whorled branches and reddish-gray rough bark. Leaves needle-like, green, fleshy, in bundles of three. Flowers arranged in cone-like structure. Male cones yellowish-brown, small, soft in nature, and short-lived. Male cones composed up of many spirally arranged microsporophylls. Female cones green when young but later on become brown, hard, and woody. Female cones long-lived and also composed up of many spirally arranged megasporophylls. Fruit woody cone with many grayish winged seeds (Fig. 3.5).

(continued)

Family Name:	Pinaceae
Medicinal Uses:	
Collection:	250 g of fresh leaves is collected in any season by men 20–40 years old; they are cut into small 1–2-in. pieces.
Recipes:	250 g of fresh leaves is ground daily for 10–15 min and 1 cup (250 mL) of water is mixed in. This mixture is filtered with a cloth and is given to patients suffering from measles. **For children,** ½ cup (125 mL) of drug (at one time) is given 3–4 times per day for 2–3 days. **For adults,** not used. The drug's taste is sour and it is green in color.
Disease Cured:	Measles.
Ethnobotanical Uses:	Leaves are used as mud roof thatching, antic racking agent in mud plasters, as fodder by goats, sheep. Wood is used for making furniture, as fuel and timber wood, and for thatching. Resin is used for removing hairs from cattle. Leaves are also used for making small brooms. Dried cones used as fuel and as decorative pieces. Seed are edible; wood is also used in handicrafts.
Phytochemicals:	Turpentine, pinene, limonene, calophony, and oleum rebinthinae [5].

3.2.3 Pinus wallichiana A.B. Jackson

Family Name:	Pinaceae
Local Name/English Name:	Kail, Saraf/Himalayan pine
Flowering Period:	February–April
Status:	Common
Part Used:	Whole tree
Habit/Habitat:	A wild evergreen tree, found on hills as self-growing plant in loamy soil in subtropical forests.
Distribution:	***Pakistan:*** Murree, Hazara, Swat, Gilliyat, Dir, and Kashmir. ***World:*** Afghanistan, Chitral eastward to W. Nepal.

(continued)

Fig. 3.6 *Pinus wallichiana* (L.)

Family Name:	Pinaceae
Description:	Trees up to 30 m tall. Bark gray, scaly. Branches whorled. Leaves acicular, in clusters of 5, bluish to gray-green. Male cones in dense clusters. Female cones 2–3 at the tips of branches, dropping, and woody; wing 2–3 times as long as the seed (Fig. 3.6).
Medicinal Uses:	
Collection:	Resin is collected by men.
Recipes:	Resin is mixed with butter and eaten before meals, which is useful for the expulsion of worms. **For children:** 1–2 tsp is given 2–3 times per day for 4–5 days. **For adults:** 2–3 tsp is used for up to 8–10 days.
Disease Cured:	Expulsion of worms.
Ethnobotanical Uses:	Wood is used for construction, i.e., doors, windows, etc., and body of trucks. Its wood is also used to make furniture and fuel.
Phytochemicals:	Turpentine contains α-pinene, undecane, dodecane, tridecane, β-pinene, Δ-3-carene, sesquiterpenes, abietic and isopimaric and ambertianic acids [4].

3.3 Section C (1) Angiosperms (Monocots)

3.3.1 Allium cepa Linn.

Family Name:	Alliaceae
Local Name/English Name:	Piaz/Onion
Flowering Period:	March–April
Status:	Commonly cultivated
Parts Used:	Bulb and leaves
Habit/Habitat:	A perennial cultivated herb, cultivated in loamy soil along with *Allium sativum*.
Distribution:	***Pakistan:*** Cultivated throughout the country. ***World:*** Cosmopolitan in distribution, found on the main islands of Indonesia, in Japan, Korea, Taiwan, Malaysia, Thailand, and Bangladesh.
Description:	A perennial cultivated herb. Stem underground bulb, with inner fleshy and outer dry scales, white to brown in color. Roots small adventitious. Leaves arise from the bulb, green, fleshy, hollow and cylindrical, basal, sheathing. Flowers are white, umbellate, perianth of six tepals, in two whorls. Fruit capsular (Fig. 3.7).

(continued)

Fig. 3.7 *Allium cepa* Linn

Family Name:	Alliaceae
Medicinal Uses:	
Collection:	Bulbs are collected by men, women, and children 12–40 years old, in summer (May–July); cleaned, washed in water 2–3 times, then dried in sunlight for 6–7 days, and stored in cotton sacks or baskets for further use.
Recipes:	**(a)** 2 or 3 bulbs (250 g) are ground daily for 8–10 min and 1 cup (250 mL) of water is mixed in. This juice is given to patients suffering from cholera. **For children,** ½ cup (125 mL) of juice (at one time) is given 3–4 times per day for 1–2 days. **For adults**, 1 cup (250 mL) of juice (at one time) is given 3–4 times per day for 1–2 days. **(b)** 60 g of fresh bulb scales is dried in sunlight for 4–5 h and then ground for 4–5 min. This powder is stored in a glass or silver pot and is given to patients suffering from diarrhea and dysentery. **For children,** 1 tsp (4–6 g) of powdered drug (at one time) is mixed in ½ cup of curd and given 2–3 times per day for 1–2 days. **For adults,** 2 tsp (10–12 g) of powdered drug (at one time) is mixed in 2 cups of curd and given 3–4 times per day for 1–2 days. **(c)** ½ kg of bulbs is ground for 8–10 min daily; then 2–3 tsp (25–30 g) of salt is mixed in. This paste is then mixed with 2–3 breads and given to cattle suffering from stomach disorders and fever (*takwo*). A dose of 250 g of drug (at one time) is given 2–3 times per day for 3–4 days. **(d)** 2 to 3 scales of *Allium cepa* are slightly dipped in *Brassica campestris* (*sarson*) oil and then warmed for 1–2 min. These scales are then secured on top of skin lesions once daily for 2–3 days.
Diseases Cured:	Cholera, diarrhea, dysentery, skin diseases, pimples, skin lesions, stomach disorders, and fever (cattle).
Ethnobotanical Uses:	Bulbs and leaves are in used in chutneys, salads, and curries, as spices and condiments.
Phytochemicals:	Volatile oil, sulfur, essential oil, organic sulfur, quercetin, moisture, ether, albuminoids, carbohydrates, fiber, ash, and sugar [6].

Fig. 3.8 *Allium sativum* Linn

3.3.2 Allium sativum Linn.

Family Name:	Alliaceae
Local Name/English Name:	Thoom, Lahsan, Oga/Garlic
Flowering Period:	March–April
Status:	Commonly cultivated
Parts Used:	Bulbs and leaves
Habit/Habitat:	A perennial cultivated herb, cultivated in loamy soil along with *Allium cepa*.
Distribution:	***Pakistan:*** Cultivated throughout the country. ***World:*** Cosmopolitan in distribution, found on the main islands of Indonesia, in Japan, Korea, Taiwan, Malaysia, Thailand, and Bangladesh.
Description:	A cultivated herb. Stem underground bulb, in clusters. Bulb with 8–16 bulblets, which are white, oval in shape, and covered by white scales. Roots small and adventitious. Leaves simple, green, fleshy, hollow, cylindrical, and radial. Flowers small, white umbellate. Fruit capsular (Fig. 3.8).
Medicinal Uses:	
Collection:	Bulbs are collected by men, women, and children 12–40 years old, in summer (April–June). They are cleaned and washed in water 2–3 times and dried in sunlight for 6–7 days. They are stored in cotton sacks or baskets for further use.

(continued)

Family Name:	Alliaceae
Recipes:	**(a)** 20 to 30 bulblets of *Allium sativum* are fried in 2 cups (500 mL) of *Brassica campestris* (*sarson*) oil for 10–15 min; when the bulblets turn black, the oil is filtered with a cloth and stored in a glass bottle, to be used for earaches. Two to three drops of oil (at one time) are used, 2–3 times per day for 1–2 days. **(b)** 8 to 10 fresh bulblets are ground daily for 3–4 min. This paste is applied to skin lesions, pimples, and other skin infections. Four to 5 g of paste is applied twice daily (morning–evening) for 6–7 days.
Diseases Cured:	Earache, skin infection, skin lesions, and pimples.
Ethnobotanical Uses:	Both leaves and bulbs are used in chutneys, in cooking curries, as spices and condiments. They are aromatic, stomachic, and a flavoring agent.
Phytochemicals:	Water, fat, protein, pectin, mucilage, Na, K, Fe, Ca, S, vitamins B and C [7].

3.3.3 *Acorus calamus L.*

Family Name:	Araceae
Local Name/English Name:	Skhawaja/Sweet flag, Calamus
Flowering Period:	April–October
Status:	Common
Part Used:	Rhizome
Habit/Habitat:	Rhizomatous herb of moist habitat.
Distribution:	***Pakistan:*** Chitral, Peshawar, Kashmir, Dir, Hazara, and Swat. ***World:*** North and Central America, Europe, and Asia.
Description:	Perennial herb up to 80 cm tall. Rootstock stout, creeping with long fibrous roots from the lower surface. Stem erect, glabrous, grooved at one side, and ribbed at the opposite. Leaves linear, Spathe leaf-like, Spadix long, cylindrical, obtuse. Tepals long, oblong-obovate, slightly curved, margin membranous, surface with embedded raphides (Fig. 3.9).

(continued)

Fig. 3.9 *Acorus calamus* L

Family Name:	Araceae
Medicinal Uses:	
Collection:	Rhizomes are collected by women and children during late November to February.
Recipes:	**(a)** Dried rhizomes are ground along with table salt. This powder is mixed with butter or ghee; tablets are made and are used with milk as a tonic, especially by women with irregular menstrual cycles. **For children,** not used. **For adults,** one tablet is given twice a day for 10–20 days. **(b)** Fresh rhizomes are crushed and squeezed to get the extract. A little water is added. Both the extract and powder drug are used in colic, dyspepsia, and flatulence. **For children:** Half cup (125 mL) is used twice a day for 8–10 days. **For adults:** One cup (250 mL) thrice a day for 15–20 days.
Diseases Cured:	Irregular menstrual cycle, colic, dyspepsia, and flatulence.

(continued)

Family Name:	Araceae
Phytochemicals:	Hydrocarbon, acorin, trimethylamine asarone acorenone, beta-asarone, calamendiol, a-selinene, a-calacorene, calamusenone, camphone, and shyobunone, neutral lipids, glycolipids, phospholipids, fatty acids, and essential oil [8].

3.3.4 Arisaema utile Hook.f.ex Schott

Family Name:	Araceae
Local Name/English Name:	Adbes, Wagmiwa, Sanpdibooti/Snake lily
Flowering Period:	May–June
Status:	Common
Parts Used:	Tuber and cob seed
Habit/Habitat:	A wild herb, mostly found at shady and cold places in loamy soil.
Distribution:	***Pakistan:*** Kashmir, Hazara, Changla Gali, Kaghan, Nathia Gali, Shogran, Dana, Sharan, and Thandiani. ***World:*** Bhutan, Afghanistan, Tibet, China, and Assam and Sikkim in India.
Description:	Wild herb, 16–33 cm tall. Bulbs subglobose, depressed with fibrous roots. Leaves solitary with three leaflets, subsessile, wavy margin. Middle one is orbicular to ovate. Leaflets form an umbrella-like structure. Petiole 20–35 cm long. Spathe tube 5–8 cm long, brownish-purple, white-ribbed; style purplish, stigma subcapitate. A cob-like structure developed on tuber bears a short stalk. Berry ovoid to subglobose, broad. One to three seeds (Fig. 3.10).
Medicinal Uses:	
Collection:	3 to 4 tubers and one to two cobs (fruit) are collected by men 20–40 years old, in late summer (August–November); stored in cloth or mud pot for further use.

(continued)

Fig. 3.10 *Arisaema utile* Hook.f.ex Schott

Family Name:	Araceae
Recipes:	**(a)** 3 to 4 fresh tubers (1 kg) are added to a mud pot. The pot's mouth is covered tightly by a lid and then these tubers are roasted on a fire for 25–30 min. Next, the black roasted tubers are ground for 4–5 min. This black-colored powdered (*kakh*) is stored in a glass bottle and is given to patients suffering from asthma. **For children,** not used. **For adults,** 2–3 g of this powdered drug (at one time) is put into one dried grape and is given at night (bedtime) daily for 15–20 days. **(b)** One fresh tuber is cut with a knife into round pieces. These pieces are placed on skin lesions or pimples and held in place by a piece of cloth daily once at night (bedtime) for 1–2 days. **(c)** 1 to 2 red fruit grains are swallowed by men and women 25–45 years old one time daily for 2–3 days in case of gas trouble and stomach disorder (*baddish*). This drug is not given to children.

(continued)

Family Name:	Araceae
Diseases Cured:	Gas trouble, stomach disorder, asthma, and skin problems (lesions and pimples).
Ethnobotanical Uses:	The tuber is eaten by porcupines.
Phytochemicals:	N-alkanes, N-alkanols, stigmasterol, sitosterol, compesterol, cholesterol, choline chloride, malic acid, fatty acid, palmitic acid, oleic acid, and linoleic acid [9].

3.3.5 *Colchicum luteum Baker*

Family Name:	Colchicaceae
Local Name/English Name:	Phanphor, Ziarguly/Golden collyrium
Flowering Period:	January–April
Status:	Common
Part Used:	Corm
Habit/Habitat:	A wild perennial herb, mostly found in open places in clay along the sides of cultivated fields in association with different grasses, and in forest shrubbery.
Distribution:	***Pakistan:*** Waziristan, Chitral, Dir, Gilgit, Baltistan, Murree, Kaghan, Margalla, Poonch, Kashmir, and Hazara. ***World:*** Central Asia, Afghanistan, and Himachal Pradesh in India.
Description:	A small perennial herb. Stem underground oval-shaped corm, convex on one side and flat on the other, with small adventitious roots. Leaves simple, long, entire, lanceolate, 3–6 in number, dark green. One to three flowers, yellow, appear before leaves, 3–4 cm broad. Perianth tube 7–9 cm long. Tepals are lanceolate to oblanceolate, 2–3 cm long. Fruit capsule ovoid with many small brownish seeds (Fig. 3.11).
Medicinal Uses:	
Collection:	125 g of fresh corms is collected by men 25–40 years old, in winter (January–March). The corms are cleaned with water to remove mud, and their outer covering is removed by hand because it is dry and rough.

(continued)

Fig. 3.11 *Colchicum luteum* Baker

Family Name:	Colchicaceae
Recipes:	100 g of fresh corms is boiled in 1 cup (250 mL) of water for 10–15 min. Then the water is filtered in a filtration pot and the corms are dried in sunlight for 4–5 h. Then they are mixed in 60–70 g of *Aloe vera* (Musabbar, Alons) and 50 g of *Terminalia chebula* (Harir) and ground for 10–15 min. Thirty or forty small tablets, 5–6 g each, are made from this powder; they are stored in a glass or plastic bottle and given to patients suffering from rheumatism. **For children,** not used. **For adults,** 1 tablet (5–6 g) of drug (at one time) is given with 1 cup (250 mL) of milk or water twice daily (morning–evening) for 15–20 days.
Disease Cured:	Rheumatism.
Phytochemicals:	Colchicine, tannic, gallic acids, starch, sugar, gum, and alkaloids [10].

Fig. 3.12 *Aloe vera* L

3.3.6 Aloe vera Linn.

Family Name:	Liliaceae
Local Name/English Name:	Kunvargandel, Ghee kuvar, Elwa/Pakistani aloe
Flowering Period:	June–September
Status:	Uncommon
Part Used:	Pulp of leaves
Habit/Habitat:	A wild as well as cultivated perennial herb, mostly found on rocky places in clay along with different grasses. It is cultivated by planting suckers, which are separated from mature plants.
Distribution:	***Pakistan:*** Sind, Punjab, Hazara, Haripur, Rawalpindi, and Islamabad. ***World:*** Cultivated in different tropical regions of the world.
Description:	A small prostrate perennial herb. Roots adventitious, arise from stem. Stem underground, thick rhizome. Leaves large, dull green to red, arise from underground stem, succulent with spinose margin, and sessile. Flowers orange–red, long funnel shape, pedicellated, and in racemes (Fig. 3.12).

(continued)

Family Name:	Liliaceae
Medicinal Uses:	
Collection:	5 to 6 fresh leaves are collected by men 25–40 years old, in winter (August–March). The leaves are washed in water 1–2 times, their rind is peeled off with a knife, and the pulp is collected.
Recipes:	**(a)** A mixture of 250 g of freshly collected pulp, 250 g of *desighee* (fat), 250 g of sugar, and 250 g of flour (*sooji*) is cooked together for 25–30 min. This sweet meal (*halwa*) is given to patients suffering from rheumatism, constipation, phlegm, indigestion, back problems, and body weakness. **For children**, not used. **For adults,** one plate of *halwa* (60–70 g) is given daily at bedtime for 10–15 days.
	(b) 1 tsp (6–8 g) of fresh pulp (at one time) is applied on pimples, skin lesions, and inflamed parts of the body 2–3 times per day for 4–5 days.
Diseases Cured:	Rheumatism, constipation, phlegm, body weakness, indigestion, back problem, pimples, and skin lesions; also used to cool inflamed body parts.
Phytochemicals:	Chromanol, pteroyglutamic acid, aloe-emodin, quinone, d-glucitol, glucosamine, mono and penta saccharides, hexuronic acid, casanthranol I and II, aloetic acid, sapogenin, glucoside, hecogenin, 2-amino-2 deoxy glucose, chrysophanic acid, m-protocatechuic aldehyde, cellulose, proteinase, resins, and imidazole [11].

3.3.7 Cynodon dactylon (L.) Pers.

Family Name:	Poaceae
Local Name/English Name:	Khabbal, Dubkhabbal, Talla/Bermuda grass
Flowering Period:	June–October
Status:	Common
Part Used:	Whole plant

(continued)

Family Name:	Poaceae
Habit/Habitat:	A wild perennial herb, commonly found in open, waste places in planes as well as hilly areas in clay.
Distribution:	***Pakistan:*** Peshawar, Baltistan, Chitral, Kurram, Hazara, Kaghan, Mansehra, Rawalpindi, Murree Hills, Karachi, Tharparkar, and Baluchistan. ***World:*** Found worldwide in tropical and warm temperate regions.
Description:	Prostrate perennial herb with a slender underground rhizome. Culms slender, up to 40 cm tall. Leaf blades short and narrow, glaucous, with or without scattered hairs; ligules are short ciliolate rim. Racemes long in a single whorl. Spikelets long; glumes lanceolate; lemma silky with simple hairs (Fig. 3.13).
Medicinal Uses:	
Collection:	100 g of fresh plant materials is collected by men and women 20–50 years old, in any season (March–November), then washed and cleaned with water 2–3 times.
Recipes:	100 g of fresh plant material is ground daily for 4–5 min. This paste (*malum*) is then applied (20–30 g) on wounds and other infected parts of the body 2–3 times per day for 1–2 days.
Diseases Cured:	To control bleeding and help with wound healing.
Ethnobotanical Uses:	The plant is used as fodder for milk-producing cattle to increase their milk and butter production.
Phytochemicals:	Protein, fiber, ether, ash, nitrogen-free extract, ether extract, mineral constituents of ash and ash sol in HCl 5.60%, CaO 0.77%, P_2O_5 0.59%, Na_2 0.23%, and KO_2 2.08%; protein, fat, total carbohydrate, fiber, ash, Ca, P, Fe, K, beta-carotene, cynodin, hydrocyanic acid, and triticin [12–14].

3.3.8 *Cyperus rotundus Linn.*

Family Name:	Cyperaceae
Local Name/English Name:	Ghawa kochin, Kai, Motha/Cyperus roots
Flowering Period:	July–September

(continued)

Fig. 3.13 *Cynodon dactylon* (L.) Pers

Family Name:	Cyperaceae
Status:	Common
Part Used:	Underground stolons

(continued)

Fig. 3.14 *Cyperus rotundus* Linn

Family Name:	Cyperaceae
Habit/Habitat:	A small grass-like herb, mostly found in shady and moist places in loamy soil.
Distribution:	***Pakistan:*** Mach, Loralai, Peshawar, Tank, D. I. Khan, Hazara, Swat, Chitral, Gilgit, Lahore, Sialkot, Attock, and Kashmir. ***World:*** All warm regions of the world, including India, China, Malaysia, and Africa.
Description:	Small herb. Stem underground rhizome, bears adventitious roots. Roots have tuberous underground stolons. Leaves simple, linear, grass-like, entire, crowded near the base of stem, green. Flowers pinkish-white (Fig. 3.14).
Medicinal Uses:	
Collection:	150 g of fresh underground stolons is collected by men 25–40 years old, in summer (June–August). They are cleaned, washed 2–3 times in water, and dried in sunlight for 3–4 days.

(continued)

Family Name:	Cyperaceae
Recipes:	A mixture of 125 g of dried stolons and 60 g of sugar is ground together for 6–7 min. Then the powder is mixed in 125 g of cowghee. This drug (*halwa*, "sweet dish") is stored in a silver or plastic pot and given to patients suffering from piles and sexual debility. **For children,** not used. **For adults,** 2 tsp (12–15 g) of drug (at one time) is given daily in the morning, before breakfast, for 10–15 days.
Diseases Cured:	Piles and sexual debility.
Phytochemistry:	Volatile oil, pinene, cineole, phenols, cyperene, cyperol, cyperone, essential oil, resinous matter and starch granules, alkaloids, glycerol; linolenic, linoleic, oleic, myristic, and stearic acids [15, 16].

3.3.9 Polygonatum verticillatum All.

Family Name:	Liliaceae
Local Name/English Name:	Noor-i-alam/Whorled Solomon's seal
Flowering Period:	May–July
Status:	Common
Part Used:	Rhizomes
Habit/Habitat:	A small perennial herb found in temperate areas. **Distribution:** ***Pakistan:*** Swat, Kurram Agency, Chitral, Dir, Kashmir, Gilgit, and Baltistan. ***World:*** Southeast Tibet, W. Asia, Europe, Russia, and Afghanistan.
Description:	Rhizomes tuber-like. Stem erect, angled, glabrous. Leaves in whorls of 3–6–8, occasionally alternate near base of stem, sometimes opposite near apex, subsessile, linear. Inflorescence flowered, perianth pale yellow. Fruit berries red, becoming purple (Fig. 3.15).
Medicinal Uses:	
Collection:	½ kg of rhizomes is dug out by men and women 20–40 years old, in the winter; dried in the shade for 15–20 days.

(continued)

Fig. 3.15 *Polygonatum verticillatum* All

Family Name:	Liliaceae
Recipes:	200 g dried rhizome is ground, powder is mixed with *desighee*, and tablets are made. These tablets are used with milk as a tonic and as an aphrodisiac. **For children,** not used. **For adults,** 1–2 tablets are given along with milk for 15–20 days.
Diseases Cured:	Used as tonic and aphrodisiac.
Phytochemicals:	Saponin and alkaloids [17].

3.3.10 Sorghum halepense (L.) Pers.

Family Name:	Poaceae
Local Name/English Name:	Baru, Lamjak/Khuskus grass, Johnson grass
Flowering Period:	July–November
Status:	Common
Part Used:	Whole plant
Habit/Habitat:	A perennial wild herb, found in clay of open and waste places.
Distribution:	***Pakistan:*** Chitral, Swat, Hazara, Gilgit, Hazara, Sargodha, Rawalpindi, Jhelum, Mianwali, Sahiwal, Lahore, Multan, Hyderabad, Tharparkar, Baltistan, and Kashmir. ***World:*** Mediterranean region eastward to Kashmir and southward to Madras. It is widely grown in the temperate regions, especially in North America.

(continued)

Fig. 3.16 *Sorghum halepanse* (L.) Pers

Family Name:	Poaceae
Description:	Rhizomatous herb; culms simple or branched up 3 m. Leaf blades long and wide. Panicle loose when in flower, lanceolate to pyramidal, long and wide. Primary branches compound and ultimately bearing racemes of 1–5 spikelet pairs. Sessile spikelets elliptic, long, hairy, cream to yellow in color. Lower glume keeled above, the wings of the keel widening upward to end in minute teeth. Upper lemma is acute, two-lobed, and long. Pedicilate spikelet is often purplish (Fig. 3.16).

(continued)

Family Name:	Poaceae
Medicinal Uses:	
Collection:	200 g of fresh roots is collected by men 20–40 years old, in winter (August–December). These roots are cleaned and washed in water 2–3 times and then cut into small 1–2-in. pieces.
Recipes:	200 g of fresh roots is ground daily for 8–10 min and then 1 cup (250 mL) of water is added. This mixture is filtered with a cloth and given to dogs. One cup of drug (at one time) is given twice daily for 3–4 days.
Diseases Cured:	Canine diseases (*zarwad*) and indigestion.
Ethnobotanical Uses:	The plant is used in both fresh and dried forms as fodder for cattle. This grass is also used as cover (*ter*) of *ghara* (stock grass) to prevent water entry.
Phytochemicals:	HCN, volatile oil, sesquiterpenes, trans-α-berg-amotene, α-patchoulene, α-himachalene [18, 19].

3.3.11 Triticum aestivum Linn.

Family Name:	Poaceae
Local Name/ English Name:	Kank, Ghandum, Ghanum/Wheat
Flowering Period:	March–April
Status:	Common
Part Used:	Whole plant
Habit/Habitat:	A cultivated annual herb, cultivated in fertile loamy soil.
Distribution:	It is commonly cultivated throughout Pakistan and the world.
Description:	An annual cultivated herb. Leaf blades long and flat. Inflorescence is a spike. Spikelets solitary at the nodes of the tough rhachis, laterally compressed, flowered. The upper one or two florets are sterile; glumes subequal, asymmetrical keeled, one- or two-toothed; lemmas keeled toward the tip, scabrid; palea membranous, two-keeled (Fig. 3.17).
Medicinal Uses:	
Collection:	Dry seeds are collected by men, women, and children 12–50 years old, in summer (March–May).

(continued)

Fig. 3.17 *Triticum aestivum* Linn

Family Name:	Poaceae
Recipes:	**(a)** 10 kg of wheat grains is soaked in 20 L (2,000 mL) of water in a big silver pot for 10–15 days. Then a white milky juice is extracted from these grains by pressing on them with bare hands. This juice is strained with a cloth in another silver pot and then placed in sunlight for 5–6 days until it changes into a white powdery mass called *nashashta*. This white powdery mass is stored in cloth sacks for further use. Fifty grams of this white powdery mass (at one time) is cooked daily in 2–3 tsp (20 g) of sugar, 2–3 tsp (20 g) of *desighee* (fat), and 1–2 cups (250–500 mL) of water for 8–10 min. This sweet meal (*halwa*) is given to nursing mothers to increase milk production for 8–10 days after childbirth. **(b)** 100 g of wheat flour is mixed daily with 50 g of *Trachyspermum ammi* (*ajwain*) and 20 g of salt. Then 1–2 cups (250–500 mL) of water is added to make it into dough. The dough is placed on a hot iron plate and cooked for 8–10 min. This bread is given to cattle suffering from cold, fever, and stomach disorders (*takwo*) and to increase their milk production. One bread (150 g) is given twice daily (morning–evening) for 2–3 days.

(continued)

Family Name:	Poaceae
Diseases Cured:	Stomach disorders, fever in cattle, and to increase milk production in cattle and nursing mothers.
Ethnobotanical Uses:	The plant is used in both fresh and dried forms as fodder for cattle, goats, and sheep. The husk (*pho*) is stored in small huts (*phoharah*) and used as fodder. The *pho* is also mixed with clay plaster as an anticracking agent on stonewalls. Grain flour is a major source of food. Grains are also soaked or boiled in water and given to cattle and other livestock to increase milk production.
Phytochemicals:	Proteins, starch, fat, fiber, ash, sugar, nitrogenous substances, salts, and acids [18].

3.3.12 Zea mays Linn.

Family Name:	Poaceae
Local Name/English Name:	Maki, Makka, Makkiia/Maize
Flowering Period:	July–August
Status:	Common cultivated plant
Part Used:	Whole plant
Habit/Habitat:	A cultivated annual herb, cultivated in fertile loamy soil.
Distribution:	It is commonly cultivated throughout Pakistan and the world.
Description:	A stout, annual herb. Roots adventitious and develop from lower nodes of the stem. Stem solid and provided with nodes and internodes. Leaves simple; alternate, long, flat with distinct sheathing base. Ligules present at the junction of the lamina and sheath. Flowers of two types in different inflorescences. Staminate inflorescence panicle at the top of main axis with lateral branches. Each spikelet bears two flowers. Carpellate inflorescence a spadix commonly called a cob, which is enclosed in a number of large bracts or spathes. The carpellate inflorescences arise axillary in the axils of lower leaves on the stem (Fig. 3.18).

(continued)

Fig. 3.18 *Zea mays* Linn

Family Name:	Poaceae
Medicinal Uses:	
Collection:	2 kg of fresh stigmas (silk) is collected by men or women 20–40 years old, in summer (July–August). Stigma is used in both fresh and dried forms; dried in sunlight for 3–4 days and stored in cloth sacks for further use.
Recipes:	30–40 g of dry or fresh stigmas, 5–7 ground grains of black pepper, and 4–5 grains of *Elettaria cardamomum* (Allaichi) are boiled daily in 3 cups (750 mL) of water for 15–20 min. When 2 cups (500 mL) of water is left, the liquid is filtered with a cloth. This decoction is given to patients suffering from kidney pain, stones, and urinary disorders (insufficient urination). **For children,** not used. **For adults,** 2 cups of decoction is given once per day for 2–3 days.
Diseases Cured:	Kidney pain, kidney stones, and urinary disorders (insufficient urination).

(continued)

Family Name:	Poaceae
Ethnobotanical Uses:	Delicious bread made from its flour is part of the normal diet of the local people. Cobs are cooked in a number of ways. Cob cases are used to clean utensils. The whole plant is used as fodder for cattle, goats, and sheep, fresh in summer and dried in winter. The plant is also used as firewood when dried.
Phytochemicals:	Amino acids, vitamins, starch, allantoin, hordenine, alkaloids, adipic acids, fixed oil, maizenic acid, sugar, resin, tannin, and salts [16].

3.4 Section C (2) Angiosperms (Dicots)

3.4.1 Acacia catechu (L.F.) Wild

Family Name:	Mimosaceae
Local Name/English Name:	Khair, Katha, Kuth/Cutch tree
Flowering Period:	April–August
Status:	Common
Part Used:	Whole plant
Habit/Habitat:	A deciduous medium-sized tree, found on waste places in dry soil.
Distribution:	***Pakistan:*** Western Himalayas, Hazara, Peshawar, Rawalpindi, Margalla Hills, Swat, and Attock. ***World:*** Tropical and foothill Himalayas, India, Nepal, Sikkim, Burma, and Assam.
Description:	A medium-sized tree up to 10 m tall. Stem erect, branched, woody, brown rough bark, and hook-shaped pointed spines. Leaves compound, leaflet 30–50 pairs, green. Flowers pale yellow in axillary spikes. Fruit legume. Pods are flat and brown, each dehiscing by two valves with 5–10 seeds (Fig. 3.19).
Medicinal Uses:	
Collection:	Heart wood gum (*kath*) is not collected locally, but is purchased from the local market (20 rupees/60 g).

(continued)

Fig. 3.19 *Acacia catechu* (L.F) Willd

Family Name:	Mimosaceae
Recipes:	4–5 g of *kath* is dissolved in 2–3 tsp (15–20 mL) of water and given to patients, especially children (5–15 years old), suffering from mouth sores, diarrhea, earache, sore throat, and eye diseases, 2–3 times per day for 3–4 days.
Diseases Cured:	Mouth sores, sore throat, diarrhea, earache, and eye diseases.
Ethnobotanical Uses:	Leaves used as fodder by goat and sheep. Plant used as firewood, thatching, hedges, and fencing; for making agricultural tools (handle plows) and implements; and as timber wood. Young twigs are used as toothbrushes (*miswak*).
Phytochemicals:	Tannic acid, catechuic acid, tannin, gum, oleic, lauric, myristic, palmatic, stearic, epoxy oleic, linoleic, linolenic, and arachidic acids, hydrocarbons, fatty acids, and glycerides [20].

3.4.2 Acacia modesta Wall.

Family Name:	Mimosaceae
Local Name/English Name:	Phulahi, Palausa, Phulau/Amritsar gum
Flowering Period:	April–May
Status:	Common

(continued)

Fig. 3.20 *Acacia modesta* Wall

Family Name:	Mimosaceae
Part Used:	Whole plant
Habit/Habitat:	A medium-sized deciduous tree, it is found in waste places in dry soil.
Distribution:	***Pakistan:*** Western Himalayas, Hazara, Peshawar, Rawalpindi, Margalla Hills, Swat, Dir, Salt Range, Kashmir, and Attock. ***World:*** Tropical and foothill Himalayas, India, Nepal, Afghanistan and Burma.
Description:	A medium-sized tree up to 10 m tall. Stem erect, branched, woody with brownish-green bark. Leaves compound, leaflet 3–5 pairs, petiolate, entire, small, rounded, dark brown prickles below the leaf petiole. Flowers in spikes creamish, fragrant. Fruit are flat brown pods (Fig. 3.20).
Medicinal Uses:	
Collection:	½ kg of fresh bark and 250 g of gum are collected by men 25–40 years old, in winter (August–February). An ax is used to carve wounds in the stem in late summer (August–October); gum is collected in winter (December–February), then dried in sunlight for 2–3 days, and stored in a glass or plastic bottle for further use.

(continued)

Family Name:	Mimosaceae
Recipes:	**(a)** 20 g of dried gum is roasted daily in 2–3 tsp (15–20 g) of *desighee*, for 4–5 min; 2–3 tsp (12–15 g) of sugar is also added. This roasted gum (*chir*) is given to patients, especially women after childbirth, suffering from back pain. **For children,** not used. **For adults,** one plate (20 g) of fried gum (at one time) is given three times per day for 8–10 days. **(b)** ½ kg of fresh bark of *Acacia arabica* and 0.5 kg of fresh bark of *Acacia modesta* are boiled in 2 L (2 jugs) of water for 1½h. When 4 cups (1000 mL) of water is left, it is strained by a cloth, and 250 g of dried grains of wheat *Triticum vulgare* are added in this filtrate, which is boiled again for 10–15 min. These grains are then dried in sunlight for 5–6 h and ground for 10–15 min. This powder is stored in a glass or plastic bottle and given to patients suffering from leucorrhoea, uterovesical disorder (gleets), and sexual debility. **For children,** not used. **For adults,** 1 tsp (8–10 g) of powdered drug (at one time) is given with 1 cup (250 mL) of milk twice daily for 10–15 days.
Diseases Cured:	Leucorrhoea (female disease), sexual debility, gleets, and back pain.
Ethnobotanical Uses:	Leaves are used as fodder by goats and sheep. Wood is used for making agricultural implements and tool handles. Young twigs are used as a *miswak* (toothbrush). Branches are used for thatching and as firewood and hedges.
Phytochemicals:	α-Amyrin, betulin, octacosanal, e-sitosterol, pet-ether, alcohol, and fatty acids [20].

3.4.3 *Acacia nilotica Linn.*

Family Name:	Mimosaceae
Local Name/ English Name:	Kikar/Black babul
Flowering Period:	March–August

(continued)

Fig. 3.21 *Acacia nilotica* Linn

Family Name:	Mimosaceae
Status:	Uncommon
Part Used:	Whole plant
Habit/Habitat:	A medium-sized tree, mostly found in dry and sunny places.
Distribution:	***Pakistan:*** Hazara, Taxila, Attock, Haripur, Margalla Hills, and Salt Range. ***World:*** Egypt, Mozambique, India, Burma, and Australia.
Description:	Moderate-sized tree up to 10 m tall. Stem erect, branched, and woody, with blackish bark and reddish-brown wood. Leaves are compound, 10–20 pairs of leaflets, opposite, entire, small, rounded, white spines at the base of each leaf. Flowers are yellow. Fruit legume, dark brown, with 8–12 seeds (Fig. 3.21).
Medicinal Uses:	
Collection:	4–5 kg of fresh bark, 250 g of fresh pods, 150 g of fresh gum, 150 g of fresh leaves, 150 g of fresh flowers, and 100 g of seeds are collected by men 25–40 years old. Leaves and flowers are collected in summer (March–July), whereas bark, gum, pods, and seeds are collected in winter (November–February). To collect the gum, an ax is used to make wounds in the stem in late summer and gum is collected in late winter (December–February). All these parts may be used in both fresh and dried forms; they are dried in shade and sunlight for 5–8 days and stored in cotton sacks or plastic or glass bottles.

(continued)

Family Name:	Mimosaceae
Recipes:	**(a)** 5 kg of fresh bark of *Acacia nilotica* and 5 kg of fresh bark of *Acacia modesta* are boiled together in 2 L (2 jugs) of water for 1½h. When only 4 cups (1000 mL) of water remain, it is filtered with a piece of cloth. Next, 250 g of *Triticum vulgare* grains is mixed in this filtrate and again boiled for 15–20 min. These grains are then dried in sunlight for 5–6 h and then ground for 8–10 min. This powder is stored in a glass or plastic bottle and given to patients suffering from gleet, leucorrhoea, and sexual debility. **For children,** not used. **For adults,** 1–2 tsp (10–12 g) of powdered drug (at one time) is given with 1 cup of milk (250 mL), twice daily (morning–evening) for 10–15 days. **(b)** 60–70 g of dried bark, 20–25 g of dried leaves, 40–50 g of dried gum, 20–25 g of dried flowers, and 20–35 g of dried pods are ground together for 10–15 min. This powder is stored in a glass or plastic bottle and given to patients suffering from diarrhea, dysentery, tuberculosis, leucorrhoea, gleet, cough, phlegm, or sexual debility and to heal wounds. **For children,** ½ tsp (3–4 g) of powdered drug (at one time) is given with 1 cup (250 mL) of milk 2–3 times per day for 2–3 days. **For adults,** 1 tsp (7–8 g) of powdered drug (at one time) is given with 1 glass of milk 2–3 times per day for 8–10 days. The drug's taste is bitter and its color is brownish. **To heal wounds,** 2–3 g of powdered drug (at one time) is sprinkled on a wound three times per day for 3–4 days.
Diseases Cured:	Diarrhea, dysentery, cough, tuberculosis, phlegm, gleet, rheumatism, leucorrhoea, gleet, and sexual debility; also used to promote wound healing.
Ethnobotanical Uses:	Leaves used as fodder by goats and sheep; twigs used as toothbrush by children; wood used as firewood when dry.
Phytochemicals:	Tannins, octacosanol, B-amyrin, B-sitosterol, butelin quercestin, epicatechol gallate, gallic acid, D-catechol, epicatechol, dimer of D-catechol, protocatechuic acid, leucoiyanidin, 3-glucoside, isoquercitrin, leucocynaidin, B-oxaly diamine propionic acid, fatty acid, amino acid, 1-arabinose, catechol, galactan, glactoaraban, n-acetyld-jenkolic acid, pentosan, and saponin [20].

3.4.4 *Achyranthes aspera Linn.*

Family Name:	Amranthaceae
Local Name/English Name:	Puth kanda, Kutri/Prickly chaff flowers
Flowering Period:	March–October
Status:	Common
Parts Used:	Fruit and roots
Habit/Habitat:	Wild annual herb, mostly grows in clay in waste places, along roads in association with different grasses.
Distribution:	***Pakistan:*** Sind, Balouchistan, Waziristan, Peshawar, Swat, Hazara, Multan, Rawalpindi, Jhelum, and Kashmir. ***World:*** Tropical regions of the world, including India, China, Japan, Indonesia, New Guinea, tropical Africa, and West Indies.
Description:	A small herb up to 50 cm tall. Stem erect, branched, and herbaceous above, woody below, light green. Leaves simple, oval-shaped, long, entire, green, mature, have hairs on under surface. Flowers small, numerous, prickly, greenish to white. Fruit oblong or oval-shaped with brown seed (Fig. 3.22).
Medicinal Uses:	
Collection:	200 g of dried prickly fruit and 100 g of fresh roots are collected by men and women 20–40 years old, in winter (mostly October–February). Roots are cleaned and washed once or twice with water.
Recipes:	**(a)** 200 g of dried fruit is placed on a hot iron plate, covered with another silver or iron lid, and burned for 5–10 min. This black-colored ash is stored in a glass or plastic bottle and is given to patients suffering from cough and asthma. **For children,** 3–4 g of ash powder (at one time) is mixed with 1 tsp (5–6 g) of honey and given 2–3 times per day for 8–10 days. **For adults,** 2–10 g of ash powder (at one time) is given 2–3 times per day for 8–10 days.

(continued)

Fig. 3.22 *Achyranthes aspera* Linn

Family Name:	Amranthaceae
	(b) 100 g of fresh roots of *Achyranthes aspera,* 100 g of fresh roots of *Boerhavia procumbens,* and 8–10 grains of "black piper" are ground together for 4–5 min. This paste-like powder is then added into 2–3 cups (500–750 mL) of water, boiled for 4–5 min, and then filtered with a cloth. This decoction is given to patients suffering from kidney problem (stone). **For children,** not used. **For adults,** 1 cup (250 mL) of decoction (at one time) is given 2–3 times per day for 4–5 days.
Diseases Cured:	Cough, asthma, and kidney stone.
Ethnobotanical Uses:	The plant is also grazed by cattle.
Phytochemicals:	Potash, saponin, oleanic acid, achyranthine, N-methyl pyrrolidine-3-carboxylic acid, betaine, and vitamin C [21, 22].

Fig. 3.23 *Ajuga bracteosa* Wall. ex Benth

3.4.5 Ajuga bracteosa Wall., ex Benth.

Family Name:	Lamiaceae
Local Name/English Name:	Kauri booti, Tarumbra/Agriculture weed
Flowering Period:	March–September
Status:	Common
Part Used:	Whole plant
Habit/Habitat:	A perennial pubescent herb, grows in sandy and clay soil in waste places.
Distribution:	***Pakistan:*** Hazara, Waziristan, Balouchistan, Dir, Kurram, Swat, Rawalpindi, Islamabad, Khudhab, and Kashmir. ***World:*** Afghanistan, Pakistan, Kashmir, Himalayas to Bhutan, Burma, China, and Malaysia.
Description:	A small prostrate, pubescent herb up to 3–4 in. tall. Stem short, reduced herbaceous, hairy. Leaves simple, opposite, petiolate, radical, toothed, dark green above and reddish green under surface, with leafy bracts. Flowers small, pale white corolla with pinkish streaks, in clusters (Fig. 3.23).

(continued)

Family Name:	Lamiaceae
Medicinal Uses:	
Collection:	150 g of fresh plant material is collected by men and women 20–40 years old. Then it is cleaned, washed in water 2–3 times, and cut into small 1–2-in. pieces.
Method:	150 g of fresh plant material is ground daily for 5–6 min and 1 cup (250 mL) of water is added. This mixture is filtered with cloth and given to patients suffering from blood diseases, mouth sores, earache, eye diseases, pimples, skin lesions, throat pain, and body inflammation. **For children,** ½ cup of extract (at one time) is given once daily before breakfast for 4–5 days. **For adults,** 1–2 cups (250–500 mL) of the drug (at one time) is given once daily before breakfast for 10–15 days.
Diseases Cured:	Mouth sores, throat pain, blood purification, body inflammation, earache, eye diseases, skin lesions, and pimples.
Phytochemicals:	Ceryl alcohol, B-sitosterol, a-sitosterol, cerotic, palmatic, oleic and linoleic acids, glucose, arabinose, thamnose, phenolic acid, natural bitter compounds, resins, iridoid glycosides, alkaloids, phytol, phytosterols, diterpenoids, triterpenoids, and unidentified compound of formula $C_{49}H_{82}O$ [9].

3.4.6 Amaranthus viridis Linn.

Family Name:	Amaranthaceae
Local Name/English Name:	Ghinar, Chalwesy, Chanlai/Green amaranth
Flowering Period:	July–October
Status:	Common
Parts Used:	Leaves and seeds
Habit/Habitat:	Common annual herb, mostly grows as weed in maize fields in clay loam.
Distribution:	***Pakistan:*** Sind, Balouchistan, NWFP, Punjab, and Hazara. ***World:*** Tropical and subtropical regions, including tropical Africa, China, India, Japan, Indonesia, New Guinea, and West Indies.

(continued)

Fig. 3.24 *Amaranthus viridus* Linn

Family Name:	Amaranthaceae
Description:	An annual herb up to 1 m tall. Stem erect, branched, herbaceous, and green. Leaves simple, alternate, petiolate, entire, lanceolate, dark green. Flowers are small, numerous, greenish. Fruit capsule with rounded dark brown to black seeds (Fig. 3.24).
Medicinal Uses:	
Collection:	250 g of seed is collected by men and women 20–40 years old, in summer (September–November). These seeds are dried in sunlight for 1–2 days and then stored in cloth or a plastic or glass bottle for further use.
Recipes:	200 g of dried seeds of *Amaranthus viridis*, 200 g of dried fruit of *Ficus carica,* and 150 g of sugar are mixed and ground together for 5–7 min. This powder-like material is then stored in a plastic or glass bottle and given to patients suffering from vision problems. **For children,** not used. **For adults,** 2–3 tsp (15–20 g) of powdered drug (at one time) is given with 1 cup (250 mL) of water twice daily (morning–evening) for 15–20 days.

(continued)

Family Name:	Amaranthaceae
Disease Cured:	To improve eyesight.
Ethnobotanical Uses:	Fresh leaves are cooked as spinach (*sag*) and also used as fodder for cattle.
Phytochemicals:	Moisture, ether, albuminoside, carbohydrates, woody fiber, ash, rutin, and quercetine [23, 24].

3.4.7 Argyrolobium roseum (Camb.) Jaub. & Spach

Family Name:	Papilionaceae
Local Name/ English Name:	Makhani booti/Roseum
Flowering Period:	April–October
Status:	Rare
Part Used:	Whole plant
Habit/Habitat:	Herb, mostly grows in waste places in dry sandy clay.
Distribution:	***Pakistan:*** Kashmir, Peshawar, Balakot, Abbottabad, Haripur, Punjab, Jhelum, Rawalpindi, Islamabad, Murree, and Balouchistan. ***World:*** Iraq, Iran, Afghanistan, India, and Nepal.
Description:	*Argyrolobium roseum* is a small prostrate herb. Stem is prostrate, branched, herbaceous and green. Leaves compound with leaflets, trifoliate, obovate, rounded at the apex and tapering at the base, smooth, shiny, whitish-green. Flowers small, pinkish-red. Fruit narrow, compressed, long pod with numerous black seeds (Fig. 3.25).
Medicinal Uses:	
Collection:	125 g of fresh plant material is collected by men and women 20–40 years old, daily on summer mornings (May–August), then cleaned and washed 2–3 times in water.
Recipes:	125 g of fresh plant material is ground for 8–10 min. Then 1 cup (250 mL) of cold water is added; the mixture is filtered with a cloth; given to patients suffering from liver, stomach, bladder inflammation, and sexual debility. **For children,** not used. **For adults,** 1 cup (250 mL) of drug (at one time) is given once daily before breakfast for 10–15 days. **Precaution:** Use only in the summer season. Do not use diluted milk (*lasi*), curd, meat, and pepper during treatment.

(continued)

Fig. 3.25 *Argyrolobium roseum* (Camb.) Jaub & Spach

Family Name:	Papilionaceae
Diseases Cured:	Liver stomach and bladder inflammation, gleets, sexual debility, skin diseases, skin lesions, and pimples.
Phytochemicals:	Vitexin and D-pinitol [25].

3.4.8 Berberis lycium Royle

Family Name:	Bereridaceae
Local Name English Name:	Sumbal, Kashmal, Kwaray, Chotra/Barberry
Flowering Period:	April–June
Status:	Common
Part Used:	Whole plant
Habit/Habitat:	A medium-sized shrub, mostly grows on hills and waste places in clay.
Distribution:	***Pakistan:*** Dir, Chitral, Gilgit, Hazara, Murree, and Kashmir. ***World:*** Afghanistan, India, China, and Nepal.

(continued)

Fig. 3.26 *Berberis lycium* Royle

Family Name:	Bereridaceae
Description:	A spiny shrub about 6 ft tall. Stem erect, branched, herbaceous above and woody below, with whitish-gray bark. Leaves simple, lanceolate, margin spiny-toothed, bright green above and pale green below, with 1–3 branched spines at the base. Flowers pale yellow. Fruit oval-shaped berry (Fig. 3.26).
Medicinal Uses:	
Collection:	5 kg of fresh roots, 300 g of fresh bark, and 200 g of fresh leaves are collected by men and women 20–40 years old. Bark and roots are collected in winter (October–March) and leaves in summer (March–August). Leaves are used in fresh form, while bark and roots are used in both fresh and dried forms. They are dried in sunlight for 7–8 days and stored in cloth for further use.
Recipes:	**(a)** 200 g of fresh leaves is boiled daily in 3 cups (750 mL) of water for 15–20 min; when 2 cups (500 mL) of water is left, then the liquid is filtered with a piece of cloth. This decoction is given to patients suffering from jaundice. **For children,** 2–3 tsp (12–15 mL) of decoction (at one time) is given 2–3 times per day for 4–5 days. **For adults,** ½ cup (125 mL) of decoction (at one time) is given 3–4 times per day for 8–10 days.

(continued)

Family Name:	Bereridaceae
	(b) 15–20 g of dried or fresh bark is soaked at night in 1 cup (250 mL) of water. The next morning it is filtered with a piece of cloth; this extract is then given to patients suffering from dysentery, eye diseases (*phora*), mouth sores, toothache, earache, throat pain, skin diseases, liver disorders, and intestinal inflammation and to purify blood. **For children,** ½–1 cup (125–250 mL) of extract (at one time) is given before breakfast for 7–8 days. **For adults,** 1 cup (250 mL) of extract (at one time) is given before breakfast for 10–15 days.
	(c) 250 g of dried bark is ground for 8–10 min. This powder is stored in a glass or plastic bottle and is used for rheumatism, bone fracture, injury, and to heal wounds. One to two teaspoons (10–12 g) of this powder (at one time) is sprinkled on injuries or wounds 2–3 times per day for 5–6 days. Twenty grams of this powder is mixed with 2–3 tsp (15–20 mL) of Brassica oil, warmed slightly for 1–2 min. This paste (*malum*) is then applied to joints and fractures once daily for 8–10 days.
	(d) 5 kg of fresh roots is washed 2–3 times in water and then cut into small 2–3-in. pieces. Then these roots are boiled in 15–25 cups (8 L) of water for 4–6 h. When only 3–4 cups (750–1000 mL) of water remains, it is filtered with a piece of cloth. This decoction is mixed with 1 kg of sugar, 1 kg of flour (*sooji*), and 300–400 g of ghee (*desi*) and cooked for 25–30 min. This sweet meal (*halwa*) is then put in an airtight plastic or silver pot and given to patients suffering from rheumatism, internal wounds (ulcer), internal injuries, and back problems. **For children,** not used. **For adults**, 100–150 g of this sweet meal (at one time) is given daily before breakfast for 20–25 days.
Diseases Cured:	Jaundice, rheumatism, injury, bone fracture, toothache, earache, eye diseases, skin diseases, mouth sores, throat pains, back pain, dysentery and liver disorder; to heal wounds; to purify blood; to calm intestines.
Ethnobotanical Uses:	Fruits are edible, leaves are used as fodder for goats and sheep, plant used for hedges and fencing, and also as firewood when dry; spines are used by young girls to pierce their ears and noses for ornaments.
Phytochemicals:	Alkaloids umbellatine, barberin, barbamine, starch grains, and tannins [26].

Fig. 3.27 *Bergenia cilita* (Haw) Stemb

3.4.9 Bergenia ciliata (Haw.) Sternb.

Family Name:	Saxifragaceae
Local Name/ English Name:	Batpia, Zakham-e-Hayat/Stone breaker
Flowering Period:	March–May
Status:	Common
Part Used:	Rhizome
Habit/Habitat:	A perennial rhizomatous herb, found on rocky slope crevices in black loamy soil.
Distribution:	***Pakistan:*** Hazara, Murree Hills, and Kaghan areas. ***World:*** Afghanistan, India, Bhutan, southernTibet, and Assam.
Description:	A small perennial rhizomatous herb with thick rootstock. Stem short, thick, fleshy, covered with persistent leaf bases, underground and dark brown. Leaves are simple, large, basal, rounded, leathery, petiolate, dark green, in autumn turning red with entire hairy margin. Flowers pinkish or pinkish-white, more than one in racemes with five sepals and petals. Fruit rounded, many-seeded capsule (Fig. 3.27).
Medicinal Uses:	
Collection:	1 kg of rhizome is collected by men 20–40 years old, in winter (August–January), and then cleaned and cut into small 1–2-in. pieces, dried in sunlight for 8–10 days, and then stored in cotton sacks or baskets for further use.

(continued)

Family Name:	Saxifragaceae
Recipes:	**(a)** 150 g of dried rhizome is ground for 8–10 min. This powder is stored in a glass bottle and is given to patients suffering from ulcer, dysentery, back pain, internal and external wounds, and piles. **For children,** 1 tsp (6–8 g) of powdered drug (at one time) is given twice daily (before breakfast and at bedtime) with 1 cup of milk for 4–8 days. **For adults,** 2 tsp (10–15 g) of powdered drug (at one time) is given twice daily (before breakfast and at bedtime) with 1 cup (250 mL) of milk for 8–10 days. **(b)** 150 g of dried rhizome is ground first for 6–7 min and then boiled in 1 cup (250 mL) of milk; 20–30 g of gum of *Acacia arabica* is also added. After 15–20 min of boiling, a paste (*halwa*) is formed, which is given to patients suffering from ulcer, back problems, and body pain. **For children,** not used. **For adults,** 2–3 tsp (15–20 g) of drug (at one time) is given twice daily (morning–evening) for 8–10 days. **(c)** 150 g of fresh bark is ground for 8–10 min daily; 1 cup (250 mL) of water is also added. Then it is filtered with a cloth and given to patients suffering from piles. **For children,** not used. **For adults,** 1 cup (250 mL) of drug (at one time) is given once daily, before breakfast, for 10–15 days.
Diseases Cured:	Ulcer, back pain, piles, dysentery, and external or internal wounds.
Phytochemicals:	Tannic acid, gallic acid, starch, mineral salts, metarbin, albumen, glucose, mucilage, wax, ash, (+) – catechine and afzelechin [27–29].

3.4.10 *Boerhaavia procumbens Banks ex Roxb.*

Family Name:	Nyctaginaceae
Local Name/ English Name:	Itsit/Red Hogweed
Flowering Period:	July–September
Status:	Common
Part Used:	Whole plant
Habit/Habitat:	A small herb, found in waste places in clay loamy soil.

(continued)

Fig. 3.28 *Boerhaavia procumbens* Benks. ex Roxb

Family Name:	Nyctaginaceae
Distribution:	***Pakistan:*** Hazara, Swat, Malakan, Murree Hills, Margalla Hills, Attock, Rawalpindi, and Kaghan areas. ***World:*** tropical Africa, tropical and subtropical Asia and America.
Description:	A tall prostrate, diffuse herb up to 1 m in length. Root large, fusiform; stems usually several, prostrate or ascending, branched, slender, cylindric, thickened at the nodes, minutely pubescent or nearly glabrous. Leaves long, broadly ovate or suborbicular, rounded at the apex, green and glabrous above, green or white beneath, the margins entire. Flowers small and pink (Fig. 3.28).
Medicinal Uses:	
Collection:	125 g of fresh plant material is collected daily by men and women 20–40 years old, in summer (April–September). Then it is cleaned and washed in water 2–3 times and cut into small 1–2-in. pieces.
Recipes:	125 g of fresh plant material is boiled daily in 3 cups (750 mL) of water, with 2–3 tsp (20–25 g) of sugar added to it. When 2 cups (500 mL) of water is left, the water is filtered. Given to patients suffering from jaundice, constipation, and internal inflammation. **For children,** 2 tsp (15 mL) of decoction (at one time) is given 2–3 times per day for 8–10 days. **For adults,** ½ cup (125 mL) of decoction (at one time) is given 2–3 times per day for 10–15 days.
Diseases Cured:	Jaundice, constipation, and internal inflammation.
Ethnobotanical Uses:	The plant is also used as fodder for cattle, goats, and sheep.
Phytochemicals:	Alkaloid punaranavine [30].

Fig. 3.29 *Bombax ceiba* Linn

3.4.11 Bombax ceiba Linn.

Family Name:	Bombacaceae
Local Name/English Name:	Dug sumbal, Sembul, Semul/Silk cotton tree
Flowering Period:	March–October
Status:	Rare
Parts Used:	Bark and wood
Habit/Habitat:	A large tree, found as a rare plant in dry clay; also planted in the plains as a roadside and garden tree.
Distribution:	***Pakistan:*** Sub-Himalayan tract from Hazara eastward. ***World:*** India, Bhutan, and South China.
Description:	A large tree up to 20 m tall. Stem erect, branched, woody, gray bark armed with prickles. Leaves compound, 5–7 leaflets, green, entire, elliptic to lanceolate with long stalks. Flowers are large, red fleshy, cup-shaped, appearing on branches before leaves. Nectariferous flowers attractive to birds. Fruit woody, capsule, dehiscing by five valves (Fig. 3.29).

(continued)

Family Name:	Bombacaceae
Medicinal Uses:	
Collection:	1 kg of fresh bark is collected by men 25–40 years old, in winter (November–February), and is used in both fresh and dried forms; dried in shade for 5–6 days; stored in cloth or a glass bottle for further use.
Recipes:	**(a)** 150 g of dried bark of *Bombax ceiba* and 150 g of dried bark of *Acacia modesta* are ground together for 8–10 min. This powdered drug is stored in a glass or plastic bottle and given to patients suffering from body weakness or rheumatism and to increase muscle strength. **For adults,** 1–2 tsp (10–12 g) of powdered drug (at one time) is given with 1 cup (250 mL) of water twice a day (morning–evening) for 14–15 days. OR 10–15 g of powdered drug is roasted daily in 5–6 g of ghee (*desighee*) and 5–6 g of sugar is mixed in. This sweet meal (*halwa*) is given to patients in the morning for 14–15 days. OR 1–2 tsp (10–15 g) of powder is mixed with 5–6 g of sugar and then dissolved in 1 cup (250 mL) of water. This syrup is given to patients daily in the morning for 12–14 days. **For children,** not used. **(b)** 10–15 g of fresh bark is ground daily for 3–4 min. This paste is then applied on pimples, skin lesions, and sore joints 1–2 times per day for 2–3 days.
Diseases Cured:	Body weakness, rheumatism, low muscle strength, skin diseases (lesions and pimples).
Ethnobotanical Uses:	The wood of the plant is used to make furniture and sometimes as timber and fuel.
Phytochemicals:	Drying oil, tannic and gallic acids, 4-C-B-D-glucopyranosyl-1,3,6,8-tetrahydroxy-7-o-(4″-hydroxybenzoyl)-9H-xanthen-9-one (I) 2-C-B-D-glucopyranosyl-1,6,7-trihydroxy-3-o(4″-hydroxybenzoyl)-9H-xanthen-9-one (II) 4-C-B-D-glucopyranosyl-1,6,8-trihydroxy-3,7-di-o-(4″-hydroxybenzoyl)-9H-xanthen-9-one (III), and mangiferin [27, 31].

Fig. 3.30 *Brassica compestris* Linn

3.4.12 Brassica campestris Linn.

Family Name:	Brassicaceae
Local Name/English Name:	Sarsoon, Wirai, Sarion/Mustard
Flowering Period:	November–April
Status:	Common cultivated plant
Part Used:	Whole plant
Habit/Habitat:	A cultivated plant, mostly cultivated in or alongside wheat fields.
Distribution:	***Pakistan:*** Cultivated in all four provinces. ***World:*** Widely cultivated throughout the world, including Europe, Central Asia, and the Mediterranean region.
Description:	A small herb up to 1 m tall. Roots cylindrical. Stem reduced at first but later becomes erect, herbaceous, branched hairy, green. Leaves simple, alternate, hairy, lobed; terminal lobe is rounded and lateral lobes are smaller, reduced in size toward the base, and sessile. Flowers yellow, in racemes. Fruit green to yellow pods (Siliqua) with numerous smaller seeds (Fig. 3.30).

(continued)

Family Name:	Brassicaceae
Medicinal Uses:	
Collection:	2 kg of fresh pods is collected by men, women, and children 12–40 years old. Fruit (pods) are collected in spring (March–April) and seeds are collected in summer (April–June).
Recipes:	2 kg of fresh pods is first ground for 8–10 min. These crushed pods are then spread on a piece of cloth and placed in sunlight for 7–8 h. After drying, the crushed pods are removed from the cloth and again ground for 3–4 min. This powder (*pakhi*) is then stored in a glass bottle and given to patients suffering from leucorrhoea, body weakness, menstrual disorder, gleets, and internal pain. **For children,** not used. **For adults,** 1–2 tsp (10–12 g) of this powder drug (at one time) is given with 1 cup of milk twice daily (morning–evening) for 15–20 days.
Diseases Cured:	Leucorrhoea, menstrual disorder, gleets, body weakness, and internal pain.
Ethnobotanical Uses:	Leaves and young stem are cooked as spinach (*sag*). Whole plant is given as fodder to milk-producing cattle to increase milk production. Brassica oil is used in cooking and during body and hair massages. Seed cakes, locally known as *khal*, are also used as delicious fodder for milk-producing cattle.
Phytochemicals:	Seed oils yield ocolaza; green top contains potash, fixed oil, sinigrin glycosides, myrosin enzyme, erucic acid, and volatile oil [6, 16].

3.4.13 *Calotropis procera* (Wild) R. Br.

Family Name:	Asclepidaceae
Local Name/English Name:	Auk, Spulmei, Madar/Swallow-wart mudar
Flowering Period:	March–December
Status:	Rare
Part Used:	Leaves and milky latex

(continued)

Fig. 3.31 *Calotrpis procera* (Wild) R. Br

Family Name:	Asclepidaceae
Habit/Habitat:	Wild shrub, grows in dry sunny places in stony, hard clay.
Distribution:	***Pakistan:*** Sind, Lower Balouchistan, Kurram, Punjab, North-West Frontier Province, Hazara, Salt Range, and Rawalpindi district. ***World:*** Afghanistan and India.
Description:	A small, erect shrub up to 2 m tall. Stem erect with branches ascending at the base, herbaceous above, woody below, whitish green bark. Plant parts are with milky juice. Leaves simple, entire, and sessile, cottony or waxy bloom, opposite, thick, upper surface green and under surface whitish. Flowers whitish and violet, in terminal or axillary cymes. Fruit follicle. Follicles are green when unripe with curved tips. Seeds with long hairs (Fig. 3.31).
Medicinal Uses:	
Collection:	2–3 cups (500–750 mL) of fresh milky latex and 250 g of fresh leaves are collected by men and women 25–40 years old, in summer (April–October).

(continued)

Family Name:	Asclepidaceae
Recipes:	**(a)** 250 g of Horse nails and 200 g of dry seed of *Prunus armeniaca* are roasted in 2–4 cups of *Brassica campestris* oil for 3–4 h. When 2 cups (500 mL) of oil remains, this mixture is ground for 15–20 min. Now 1–2 cups of *Calotropis procera* milk is mixed in. This paste-like material is then stored in a glass or silver pot and applied to infectious parts of skin. **For children,** not used. **For adults,** 2–3 g of paste (*malum*) (at one time) is applied to site of infection (*chambal*) with a hen wing 2–3 times per day for 20–25 days. **(b)** 2–3 tsp (250 mL) of fresh milk (latex) is applied over the affected parts of snakebite, dog bite, and insect sting until the patient feels relief. **(c)** 200 g of semi-dried leaves is placed on a mud plate and 100 g of common salt is sprinkled over them; this is burned for 10–15 min. Then these leaves are ground for 4–5 min; the resulting powder is stored in a glass bottle and given to patients suffering from asthma and cough. The smoke is used to kill insects. **For children,** not used. **For adults,** 4–5 g of powdered drug (at one time) is given with 1 cup (250 mL) of water 2–3 times per day for 15–20 days.
Diseases Cured:	Skin infections (*chambal*), cough, asthma, dog and snakebites, and insect stings.
Phytochemicals:	Voruscharin, uscharidin, trypsin calcatin, uzarigenin, proceroside, benzoyllineolone, benzoylisolineolone, syriagenin, calotoxin, and calotropin cyanidin-3-rhamnoglucoside [32].

3.4.14 *Cannabis sativa Linn.*

Family Name:	Cannabinaceae
Local Name/English Name:	Bhang/True hemp
Flowering Period:	July–September
Status:	Common
Part Used:	Leaves

(continued)

Fig. 3.32 *Cannabis sativa* Linn

Family Name:	Cannabinaceae
Habit/Habitat:	A common wild herb, mostly grows in moist and waste places near houses and alongside roads.
Distribution:	***Pakistan:*** Common weed throughout the country. ***World:*** India, Central Asia, and tropical Africa. It is also cultivated elsewhere.
Description:	An annual herb up to 1 m tall. Stem erect, branched, herbaceous 4–8 ft, and angular. Leaves simple, alternate, petiolate, and palmate; leaf lobes are sessile, narrow at the base, and upper surface is hairy green. Flowers numerous, pale yellow, small, and drooping. Female flowers erect; perianth a single entire leaf enclosing the ovary; style thread-like. Fruit yellowish-brown, an achine, enclosed in persistent leaf and single-seeded (Fig. 3.32).

(continued)

Family Name:	Cannabinaceae
Medicinal Uses:	
Collection:	1 kg of fresh leaves is collected by men 25–40 years old, in summer (May–August). Leaves are used in both fresh and dried forms; dried in shade for 2–4 days and then stored in a cloth sack for further use.
Recipes:	**(a)** 150 g of fresh or dried leaves and 60–70 g of sugar are ground together for 7–8 min; then 1 cup (250 mL) of water is mixed into it. This mixture is then filtered with a piece of cloth and drunk by men 25–40 years old for general body inflammation, intoxication, and loss of appetite. **For children,** not used. **For adults,** 1 cup (250 mL) of filtrate is used once daily in the morning (9–11 a.m.) for 8–10 days.
	(b) ½ kg of fresh leaves of *Cannabis sativa is* ground for 4–5 min and then mixed with ½ kg of wheat flour, *Triticum aestivum.* Then 2–3 cups (500–750 mL) of water is added. This paste is given to cattle, goats, and sheep suffering from abdominal swelling and to improve hunger; 300 g of drug (at one time) is given 2–3 times per day for 2–3 days.
Diseases Cured:	Body inflammation, intoxication, loss of appetite in humans; abdominal swelling due to indigestion in cattle.
Ethnobotanical Uses:	Plant is used as fodder for cattle and also as firewood when dry.
Phytochemicals:	Volatile oil composed of cannabene, cannabine, alkaloids, cannabinone; a resin consists of cannabinol, pseudo cannabinol, cannabinin, and several terpenes [33].

3.4.15 *Carissa opaca Stapf-ex Haines*

Family Name:	Apocynaceae
Local Name/English Name:	Granda, Karaunda, Gorna/Bengal current
Flowering Period:	April–June
Status:	Common

(continued)

Fig. 3.33 *Carrisa opaca* Stapf. ex Haines

Family Name:	Apocynaceae
Part Used:	Whole plant
Habit/Habitat:	Wild spiny shrub, mostly grows in dry clay alongside *Dodonia*, *Olea,* and *Punica* species.
Distribution:	***Pakistan:*** Hazara, Rawalpindi district, Margalla Hills, and Swat. ***World:*** India, Burma, and Sri Lanka.
Description:	A spiny shrub up to 2–3 m tall. Stem erect, branched, woody, young shoot with milky juice and straight hard spines. Spines are 20–30 cm long and bark is grayish. Leaves simple, opposite, upper surface shiny, dark green, lower surface hairy and light green, oval-shaped. Flowers sweet-scented in terminal cymes, small, white. Fruit black, purple, oblong berry with milky latex (Fig. 3.33).
Medicinal Uses:	
Collection:	1 kg of fresh leaves, 1 kg of fresh fruit, and 60–70 g of fresh roots are collected by men, women, and children 12–40 years old. Fruit and roots are collected in winter (October–February), whereas leaves may be collected in any season when required. Leaves and fruit are used in fresh condition, while roots are dried in sunlight for 3–4 days and then used.

(continued)

Family Name:	Apocynaceae
Recipes:	**(a)** 1 kg of fresh leaves of *Carissa opaca* and 1 kg of fresh roots of *Segeratia brandrethina* are boiled in 3–4 jugs (4 L) of water for 1½ h. When 1–2 jugs (1 L) of water is left, then it is filtered with a cloth or filtration pot. This decoction is stored in a glass bottle and given to patients suffering from asthma, jaundice, and kidney pain. **For children,** ½ cup (125 mL) of decoction (at one time) is mixed with 2 cups (500 mL) of water and 10–15 g of sugar; then it is shaken well and given to the patient 2–3 times per day for 6–7 days. **For adults,** 1 cup (250 mL) of decoction (at one time) is mixed with 2 cups (500 mL) of water and 10–15 g of sugar; then it is shaken well and given twice daily (morning–evening) for 5–6 days. **(b)** 30 g of dried roots is ground daily for 4–5 min. This powder is then sprinkled on animals' wounds and infected sores 2–3 times per day for 3–4 days. **(c)** 1 kg of freshly collected ripe fruit is crushed by hands and milk juice is extracted. Next, 80 g of iron filings are mixed into it. This mixture is then put in a glass bottle for 20–25 days, then ground for 15–20 min, and small tablets (3–4 g each) are made. These tablets are then given to patients suffering from liver disorders and blood deficiencies. **For children,** not used. **For adults,** one tablet (at one time) is given with 1 cup (250 mL) of water or milk twice daily (morning–evening) for 8–10 days.
Diseases Cured:	Asthma, jaundice, kidney stones, liver disorders, and blood deficiencies and used for wound healing in animals.
Ethnobotanical Uses:	Leaves are used as fodder for goats and sheep. Fruit is edible. Also used as firewood and as hedges and fencing.
Phytochemicals:	Phenols and flavonoids, palmitic acid, benzyl salicylate, benzyl benzoate, and (*E*, *E*)-α-farnesene [34, 35].

Fig. 3.34 *Cassia fistula* Linn

3.4.16 Cassia fistula Linn.

Family Name:	Caesalpinaceae
Local Name/English Name:	Kinjal/Golden shower tree
Flowering Period:	March–June
Status:	Uncommon
Part Used:	Whole plant (pods and seed)
Habit/Habitat:	A medium-sized tree grown on dry and sunny land.
Distribution:	***Pakistan:*** Hazara, Rawalpindi, Islamabad, Margalla Hills, and Haripur. ***World:*** Throughout the greater part of India, Burma, and Ceylon.
Description:	A medium-sized evergreen tree up to 15 m tall. Stem erect, branched woody, with dark gray bark. Leaves compound, alternate. Flowers numerous, bright yellow in clusters. Fruit legume, cylindrical pod with 10–15 flat dark brown seeds (Fig. 3.34).

(continued)

Family Name:	Caesalpinaceae
Medicinal Uses:	
Collection:	5 kg of dried pods are collected by men and children 14–40 years old, in winter (January–May). These pods are broken, and the internal pulp and seeds are collected. These seeds are stored in glass or plastic bottles or in cotton sacks for further use.
Recipes:	15–20 seeds (8–10 g) are boiled daily in 1 cup (250 mL) of milk; 1–2 tsp (10–12 g) of sugar is also added, and the mixture is boiled for 8–10 min. Then it is filtered with a cloth and given to patients suffering from constipation and stomach disorders. **For children,** ½ cup (125 mL) of decoction (at one time) is given once daily for 3–4 days. **For adults,** 1–2 cups (250–500 mL) of decoction (at one time) is given once daily for 8–10 days.
Diseases Cured:	Constipation and stomach disorders.
Ethnobotanical Uses:	Wood is used for making light furniture, agricultural implements, and tool handles, and as mud roof thatching and fuel. Pods are sold for cash.
Phytochemicals:	Anthraquinone, tannins, pholpaghenes, oxy anthraquinone, resin, volatile oil, wax, resin, anthraquinones, flavonoids, and flavan-3-ol derivatives [30, 36].

3.4.17 *Cedrela toona Roxb. ex Wild*

Family Name:	Meliaceae
Local Name/ English Name:	Neem, Guldar, Kanem/Toon tree
Flowering Period:	April–June
Status:	Rare
Parts Used:	Leaves, bark, and wood
Habit/Habitat:	A tall tree, found in damp shady ravines and on hillsides.

(continued)

Fig. 3.35 *Cedrela toona* Roxb. ex Willd

Family Name:	Meliaceae
Distribution:	***Pakistan:*** Hazara, Swat, Buner, Kaghan, Murree, Poonch, Jhelum, and Margalla Hills. ***World:*** Burma, Java, Australia, and India.
Description:	A tall tree up to 15 m in height. Stem erect, branched, and woody with dark brown bark. Leaves compound, 3–5 pairs, and entirely green, opposite, subsessile. Flowers creamy white, small, in clusters. Fruit capsule, single-seeded, and yellow (Fig. 3.35).
Medicinal Uses:	
Collection:	5 kg of fresh leaves and 1–2 kg of bark are collected by men 20–40 years old. Leaves are collected in summer (April–August) and bark is peeled off in winter (November–March) with an ax. Leaves are dried in shade for 4–5 days and bark in sunlight for 2–3 days. Both bark and leaves are stored in cotton sacks for further use.

(continued)

Family Name:	Meliaceae
Recipes:	**(a)** 250 g of dried leaves is ground for 10–15 min, and then 1–2 tsp (10–15 g) of common salt is mixed in. This powder is stored in a glass or plastic bottle and given to patients suffering from fever, diabetes, and skin diseases (allergy and pimples) and to purify the blood. **For children,** 1 tsp (4–6 g) of powdered drug (at one time) is given with 1 cup (250 mL) of water once daily, in the morning, for 10–15 days. **For adults,** 2–3 tsp (12–15 g) of powdered drug (at one time) is given with 1 cup (250 mL) of water once daily, in the morning, for 15–20 days. **(b)** 125 g of dried bark is ground for 10–15 min. The powdered drug is stored in a glass or plastic bottle and given to patients suffering from dysentery or ulcers and to heal wounds. **For children,** 1 tsp (4–6 g) of drug (at one time) is given with 1 cup (250 mL) of water twice daily (morning–evening) for 3–4 days. **For adults,** 2–3 tsp (12–15 g) of powdered drug (at one time) is given with 1 cup (250 mL) of water twice daily (morning–evening) for 6–7 days.
Disease Cured:	Fever, diabetes, dysentery, blood diseases, skin diseases (allergy and pimples), ulcers, and wound healing.
Ethnobotanical Uses:	Leaves are used as fodder by goats and sheep. Wood is used for making furniture, in construction, and as fuelwood.
Phytochemicals:	Resin, gum, nyctanthin, flavones, glycosides, tannic acid, resin, citric acid, starch, ash; and essential oil consists of tricyclic acid, sesquiterpene, copaene, cadinene, cadinol, lactone, and cedrelone [37].

3.4.18 *Chenopodium ambrosioides* Linn.

Family Name:	Chenopodiaceae
Local Name/English Name:	Waljuin, Chandan bathwa, Bathu/Skunkweed
Flowering Period:	April–June
Status:	Common
Part Used:	Leaves
Habit/Habitat:	Annual herb, mostly grows near houses in waste places as weed in clay loam.

(continued)

Fig. 3.36 *Chenopodium ambrosioides* Linn

Family Name:	Chenopodiaceae
Distribution:	***Pakistan:*** Balouchistan, Peshawar, Dir, Chitral, Swat, Hazara, and Kashmir. ***World:*** Widely distributed in the world, introduced into America from elsewhere and naturalized in the wild.
Description:	An erect herb up to 1 m tall. Stem erect, branched, herbaceous, and green. Leaves simple, alternate, petiolate, toothed, and green. Flowers small, numerous, yellowish-green in cymose clusters forming axillary spikes or long terminal panicles. Seeds small, numerous, rounded (Fig. 3.36).
Medicinal Uses:	
Collection:	150 g of fresh leaves is collected by men and women 20–40 years old, in summer (April–July).
Recipes:	150 g of fresh leaves is boiled daily in ½L of water for 15–20 min; when 3 cups (750 mL) of water remains, the water is filtered with a piece of cloth; given to patients suffering from piles, gas trouble, stomach griping, or indigestion. **For children,** not used. **For adults,** 1½ cup (375 mL) of decoction (at one time) are given twice daily (morning–evening) for 10–15 days.

(continued)

Family Name:	Chenopodiaceae
Diseases Cured:	Piles, gas trouble, indigestion, and grippe.
Phytochemicals:	Essential oils, saponin, vitamin C, magnesium phosphate, ascaridol, ascaridole, dimethyloxide, safrle, ascaridole glycol, butyric acid, p-cymene, L-limonene, l-isolimonene, and d-camphor [15, 30].

3.4.19 *Cichorium intybus Linn.*

Family Name:	Asteraceae
Local Name/English Name:	Kasni booti, Kashni, Hand/Wild chicory
Flowering Period:	June–September
Status:	Common
Part Used:	Whole plant
Habit/Habitat:	A perennial herb, grows as a common weed in waste places of cultivated and grassy fields in clay loam.
Distribution:	***Pakistan:*** Balouchistan, Waziristan, Kurram, Peshawar, Swat, Hazara, Astor, Gilgit, Baltistan, Murree, Rawalpindi, Islamabad, and Kashmir. ***World:*** Europe, West Asia, Afghanistan, Iran, India, and Nepal.
Description:	A perennial suberect to spreading herb up to 2 ft tall. Stem erect, branched, herbaceous, and green. Leaves simple, green, toothed, lanceolate, pinnatifid, deeply divided, crowded at the base, and spirally arranged on stem. Flowers bright blue, blue purple, or pinkish, in terminal or axillary head, pappus absent or scaly. Fruit brownish to black with ovoid seeds (Fig. 3.37).
Medicinal Uses:	
Collection:	250 g of fresh plant is collected by men and women 20–45 years old, in summer (April–September). It is washed in water 2–3 times and cut into small 1–2-in. pieces.

(continued)

Fig. 3.37 *Cichorium intybus* Linn

Family Name:	Asteraceae
Recipes:	250 g of fresh plant material and 100 g of sugar are boiled in 2 cups (500 mL) of water for 20–25 min; when 1–2 cups (250–500 mL) of water is left, the water is strained with a piece of cloth. This decoction is stored in a glass bottle and given to patients suffering from fever, jaundice, gas trouble, stomach disorders, and body swelling. **For children,** 1–2 tsp (10 mL) of decoction (at one time) is mixed with 1 cup (250 mL) of water and given 2–3 times per day for 4–5 days. **For adults,** 3–4 tsp (25 mL) of decoction (at one time) is mixed in 1 cup (250 mL) of water and given 2–3 times per day for 6–7 days.
Diseases Cured:	Jaundice, common fever, stomach disorder, body swelling, and gas trouble.

(continued)

Family Name:	Asteraceae
Ethnobotanical Uses:	The leaves are cooked like spinach (*sag*). The plant is also used as fodder for goats, sheep, and cattle.
Phytochemicals:	Gummy water, cellulose, inulin, fiber, ash, glycoside, stearin, mannites, tartaric acid, betaine, choline, lactones, esculine, esculetin, cichoriin, umbelliferone, scopoletin, 6,7-dihydroxy caumarin, volatile matter anthocynin (Lvs), fatty acid, methyl esters, vanillin, 5-hydroxy methyl 2-furfural, 2-acetylpyrole, furfural, phenyl acetic acid, 2-(5-hydroxy methyl 2-formyl pyrol-1-yl)-3-methyl-3-pentanoic lactone, and phenyl acetaldehyde [38].

3.4.20 Cissampelos pariera L.

Family Name:	Menispermaceae
Local Name/English Name:	Ghore summi, Phuldhari/False pareiar brava
Flowering Period:	July–September
Status:	Common
Part Used:	Whole plant
Habit/Habitat:	A small climbing herb, mostly found on cliffs and as a climber on other plants.
Distribution:	***Pakistan:*** Hazara, east of Indus River, Rawalpindi, Kashmir, Jhelum, and Margalla Hills. ***World:*** India and Sri Lanka.
Description:	A climbing herb up to 1 m in length. Stem climber, weak, branched, hairy, and green. Leaves simple, alternate, petiolate, green, hairy, and rounded like horse toes (foot) with ciliated margin. Flowers small, yellowish, and in clusters (Fig. 3.38).
Medicinal Uses:	
Collection:	30–40 g of fresh leaves and young stems is collected by men 20–40 years old, daily in summer (April–August).

(continued)

Fig. 3.38 *Cissampelos pariera* Linn

Family Name:	Menispermaceae
Recipes:	30–40 g of fresh plant material is ground daily for 4–5 min; ½ cup (125 mL) of water and 2 tsp (10–12 g) of sugar are mixed in. This ground material is stored in a cup overnight and given the next morning to patients suffering from inflammation, gleets, and sexual debility. **For children,** not used. **For adults,** 3–4 tsp (20–25 g) of drug (at one time) is given once daily, before breakfast, for 8–10 days.
Diseases Cured:	Intestinal and bladder inflammation, gleets, and sexual debility.
Ethnobotanical Uses:	The plant is used as fodder by goats and sheep.
Phytochemicals:	(4-Hydroxy-3-methoxyphenyl)-7-(4-methoxyphenyl)-6-(2-hydroxy-4, 6-dimethoxybenzoyl)-furano [3, 2-g] benzopyran-4-one [39].

Fig. 3.39 *Citrus limon* (Linn.) Brum

3.4.21 Citrus limon (Linn.) Burm.

Family Name:	Rutaceae
Local Name/English Name:	Nimbu, Lemu/Lemon
Flowering Period:	June–July
Status:	Cultivated plant
Part Used:	Fruit
Habit/Habitat:	A small tree, found as cultivated plant in home gardens.
Distribution:	***Pakistan:*** Distributed in all four provinces as a cultivated plant. ***World:*** Cultivated in tropical areas of the world.
Description:	A small tree up to 2 m tall. Stem erect, branched, woody, and spiny with greenish bark. Leaves compound, alternate, and leaflets, petiolate, entire, green. Flowers white. Fruit oval, yellow with white seeds, pale, and acidic pulp (Fig. 3.39).
Medicinal Uses:	
Collection:	2 or 3 fresh fruits are collected daily by men, women, or children 12–40 years old, in late summer (September–October).

(continued)

Family Name:	Rutaceae
Recipes:	**(a)** 3 tsp (20 mL) of fresh fruit juice and 3 tsp (20 mL) of fresh *Allium cepa* juice are mixed and given to patients suffering from cholera and vomiting. **For children,** 1 tsp (5 mL) of juice (at one time) is given 2–3 times per day for 1–2 days. **For adults,** 2 tsp (10–12 mL) of juice (at one time) is given 2–3 times per day for 2–3 days. **(b)** 4 cups (1 L) of water, 4 tsp (20–40 g) of sugar, and 1 tsp (8–10 g) of green tea are boiled together daily for 8–10 min, then filtered with a cloth or filtration pot, stored in a teapot, and given to patients suffering from fever. **For children,** 1 tsp (5 mL) of fresh juice is mixed with ½ cup of decoction (green tea) (at one time) and given 2–3 times per day for 3–4 days. **For adults,** 2 tsp (12–15 mL) of fresh juice is mixed in 1 cup of decoction (at one time) and given 2–3 times per day for 8–10 days.
Diseases Cured:	Cholera, fever, and vomiting.
Ethnobotanical Uses:	Leaves are used as fodder by goats and sheep.
Phytochemicals:	Votalile oil, citric acid, limonene, citral, linalool, linalyl acetate, terpinol, cymene, coumarins, isopimpinellin, bergapten, citropten, xanthyletin, phenolic compounds, as well as vitamins, minerals, dietary fiber, essential oils, and carotenoids [30, 40].

3.4.22 *Clematis grata Wall.*

Family Name:	Ranunculaceae
Local Name/English Name:	Dhand, Birli/Clematis
Flowering Period:	August–October
Status:	Common
Parts Used:	Leaves and young stem
Habit/Habitat:	A shrubby climber, mostly found in waste places in clay loam.

(continued)

Fig. 3.40 *Clematis grata* Wall

Family Name:	Ranunculaceae
Distribution:	***Pakistan:*** Hazara, Malakand, Chitral, Swat, Drosh, Lowari Top, Salt Range, Hassan Abdal, Murree, and Kashmir. ***World:*** East Mediterranean through West and Southwest Asia, South Siberia, Mongolia, Tibet, and China.
Description:	A climbing herb up to 3–5 m in length. Stem is climber, branched, herbaceous, and green. Leaves compound, pinnate or bi-pinnate, leaflets lanceolate, often basally with 1–2 shorter lobes. Flowers yellow, purplish tinged on the outside, in lax panicles, spreading, ovate-lanceolate with curved tips. Fruit is achene, silky, hairy (Fig. 3.40).
Medicinal Uses:	
Collection:	120 g of fresh leaves and young branches is collected by men and women 20–40 years old, in summer (June–October)
Recipes:	**(a)** 120 g of fresh plant material is ground daily for 7–8 min; 50–60 g of this paste (*malum*) is applied of infected body parts once per day for 3–4 h.

(continued)

Family Name:	Ranunculaceae
	(b) 60–70 g of fresh plant material is ground daily for 7–8 min and 1 cup (250 mL) of water is added; 3–4 tsp (50–60 mL) of juice is applied (at one time) 2–3 times per day for 2–3 days.
Diseases Cured:	Skin infection (*chambal*); kills germs and worms in the wounds of cattle and dogs.
Ethnobotanical Uses:	Grazed by goats.
Phytochemicals:	Glycoside (ranunculin), from which protoanemonin, an irritant, is formed [3].

3.4.23 Convolvulus arvensis Linn.

Family Name:	Convolvulaceae
Local Name/English Name:	Liali, Khurry, Erlai/Small bind weed
Flowering Period:	February–November
Status:	Common
Part Used:	Whole plant
Habit/Habitat:	A climbing, twining herb, found as weed in wheat fields and wet places in clay loam.
Distribution:	***Pakistan:*** Common and widespread as a weed from Sindh to Baltistan. ***World:*** Tropical and temperate regions.
Description:	A small prostate or climbing herb. Stem weak, herbaceous, green, twining, and cylindrical. Leaves simple, alternate, petilate, entire, lanceolate, arrowhead-shaped with two lobes directed downward, sagittate to hastate. Flowers pale pink to pink or white, pedicellate, often with darker strips. Fruit is a capsule, with 2–4 seeds (Fig. 3.41).
Medicinal Uses:	
Collection:	100 g of fresh leaves and roots is collected by men and women 25–40 years old, daily in summer (March–July), and washed in water once or twice.

(continued)

Fig. 3.41 *Convolvulus arvensis* Linn

Family Name:	Convolvulaceae
Recipes:	100 g of fresh plant material and 30–40 g of sugar are ground together daily for 4–5 min and mixed with 1 cup (250 mL) of water. This mixture is filtered and then given to patients suffering from constipation. **For children,** 1 cup of mixture (at one time) is given at night daily for 2–3 days. **For adults,** 1 cup (250 mL) of drug (at one time) is given daily at night for 7–8 days.
Disease Cured:	Constipation.
Ethnobotanical Uses:	The fresh leaves are used as fodder for cattle and also used to remove dandruff.
Phytochemicals:	Tannin, resin, convolvulin, campesterol, amyrin, stigmasterol, B-sistosterol, n-alkanes, n-alkanols, umbelliferone, and scopoletin [32].

3.4.24 Cuscuta reflexa Roxb.

Family Name:	Cuscutaceae
Local Name/ English Name:	Niladhary, Akashbel, Banosha, Zarbuti/Doddar
Flowering Period:	January–March
Status:	Uncommon

(continued)

Fig. 3.42 *Cuscuta reflexa* Roxb

Family Name:	Cuscutaceae
Part Used:	Whole plant
Habit/Habitat:	A wild twining herb, mostly found on *Zizyphus* species, forming a dense mass of yellowish threads.
Distribution:	***Pakistan:*** Sind, Balouchistan, Karachi, Dir, Chitral, Hazara, Gilgit, Salt Range, Kashmir, Muree, and Islamabad. ***World:*** Afghanistan, Southwest China to Southeast Asia.
Description:	A twining herb. Stem weak, twiner, yellow, branched, thread-like, soft, thick, succulent with small adventitious roots. Leaves absent; plant is total parasite. Flowers white cream color. Fruit is brown, capsule with 2–4 seeds (Fig. 3.42).
Medicinal Uses:	
Collection:	½ kg of fresh plant is collected daily by men or women 20–40 years old, in any season, mostly in summer (April–November).

(continued)

Family Name:	Cuscutaceae
Recipes:	½ kg of fresh plant is cut into small 2–4-in. pieces and boiled in 4 cups (1 L) of water for 1–2 h in a mud pot covered with an airtight lid. After 1–2 h, the mud pot is opened in an airtight room and the patient's infected joints or paralyzed body part(s) are placed over the steam for 8–10 min. Then the remaining paste-like material (*halwa*) is given to patients suffering from paralysis, rheumatism, and jaundice. **For children,** not used. **For adults,** 2–3 tsp (15–20 g) of fresh drug is given once at night for 6–7 days.
Diseases Cured:	Rheumatism, paralysis, and jaundice.
Phytochemistry:	Flavonoids of the flavonol type, phenolic compounds, hydroxycinnamic acid, phenylpropanoids, and caffeic acid [41].

3.4.25 *Dalbergia sissoo Roxb.*

Family Name:	Papilionaceae
Local Name/ English Name:	Tali, Shawa/Sissoo
Flowering Period:	March–April
Status:	Uncommon
Part Used:	Whole plant
Habit/Habitat:	A tall tree, mostly found as self-growing and cultivated plant in waste places in clay loam.
Distribution:	***Pakistan:*** Found as common self-growing and wild plant throughout all four provinces. ***World:*** Central Himalaya, Afghanistan, and India.
Description:	A tall tree up to 25 m in height. Stem erect, branched, woody, and hard with yellowish-gray bark. Leaves compound, 3–5 leaflets, pale green, entire, petiolate, heart-shaped terminal; one is large and the other is smaller. Flowers pale yellow. Fruit legume, pod yellowish-green with 3–4 seeds (Fig. 3.43).
Medicinal Uses:	
Collection:	250 g of fresh bark and 100 g of fresh leaves are collected by men and women 20–40 years old. The bark is collected in winter (November–February) and dried in sunlight for 2–3 days, whereas leaves are collected in spring/summer (April–August) and used fresh.

(continued)

Fig. 3.43 *Dalbergia sissoo* Roxb

Family Name:	Papilionaceae
Recipes:	**(a)** 200 g of dried bark is ground for 8–10 min. This powder is stored in a plastic or glass battle and given to patients suffering from body pain. **For children,** not used. **For adults,** 1–2 tsp (10–12 g) of powdered drug (at one time) is given once daily, at night, with 1 cup of water or milk for 15–20 days. **(b)** 100 g of fresh leaves is boiled daily in 1 cup (250 mL) of water for 10–15 min. When 1½ cups (375 mL) of water remains, it is filtered with a cloth or a filtration pot. This decoction is given to patients suffering from diarrhea. **For children,** 1–2 tsp (10–15 mL) of decoction is given (at one time) 2–3 times per day for 2–3 days. **For adults,** ½ cup (125 mL) of decoction (at one time) is given 3 times per day for 3–4 days. **(c)** 1 kg of fresh leaves is boiled daily in 5 L of water for 20–25 min. This water is filtered with a cloth and used while washing hair, to remove dandruff, for 5–6 days.
Diseases Cured:	Body pain, diarrhea, and dandruff.
Ethnobotanical Uses:	Leaves are used as fodder for goats and sheep. Wood is used for making furniture, agricultural implements, plugs, and tool handles and as fuelwood and timber, and as thatching.

(continued)

Family Name:	Papilionaceae
Phytochemicals:	Fixed oil, fatty acid, myristic, palmatic, stearic, arachidic, oleic acids crystalline lactone, dalbergin, isodalbergin, ethers, dalbergenone, isoflavone, ether, isoflavone, glycoside, caviunin, 7-ogentiobioside isocaviunin, isoflavoneglycoside, sissotrin, phynyl-chromene, dalbergichromene, and isotectrigenin [42].

3.4.26 *Datura innoxia Miller*

Family Name:	Solanaceae
Local Name/ English Name:	Tatura, Datura/Green throne apple
Flowering Period:	May–October
Status:	Rare
Parts Used:	Seeds and leaves
Habit/Habitat:	An annual herb, found in waste places and along roadsides in clay loam.
Distribution:	***Pakistan:*** Balouchistan, Jhelum, Rawalpindi district, Islamabad, Murree Hills, and Kashmir. ***World:*** Native to tropical America; naturalized elsewhere.
Description:	A small annual herb up to 1 m tall. Stem erect, branched, herbaceous, and green. Leaves simple, alternate, entire, wavy, petiolate, and green. Flowers in axillary solitary, white, and funnel-shaped. Fruit capsule, rounded, and spiny (Fig. 3.44).
Medicinal Uses:	
Collection:	250 g of seeds and 1 kg of fresh leaves are collected by men and women 20–40 years old. Seeds are collected in winter (December–February) and are dried in sunlight for 1–2 days, whereas leaves are collected in summer (April–September) and are dried in shade for 2–3 days. Both are stored in a cotton sack or plastic bottle for further use.
Recipes:	120 g of dried leaves, 60 g of dried seeds of *Datura innoxia,* and 120 g of dried leaves of *Nicotiana tabacum* are ground for 10–15 min. This powder is stored in a plastic bottle and is used by patients suffering from cough and asthma. **For children,** not used. **For adults,** 2 tsp (10–15 g) of powdered drug (at one time) is put in a pipe and is smoked by patients for 2–3 min once or twice a day for 10–15 days.

(continued)

Fig. 3.44 *Datura innoxia* Miller

Family Name:	Solanaceae
Diseases Cured:	Cough and asthma.
Phytochmeicals:	Hyoscymine, atropine, apoatropine, belladonnine, scopolamine, resin, and daturine. Seeds contain fixed oil, ditigloyl esters of 3, 6-dihydro tropane, 3, 6, 7-trihydrotropane [26].

3.4.27 Debregeasia saeneb (Forssk.) Heper and Wood

Family Name:	Urticaceae
Local Name/ English Name:	Sandoori/Wild rhea
Flowering Period:	March–April
Status:	Common
Part Used:	Whole plant
Habit/Habitat:	An evergreen shrub, mostly found in waste places of lower hills in sandy soil.
Distribution:	***Pakistan:*** Hazara, Swat, Dir, Hassan Abdal, Salt Range, Murree Hills, Margalla Hills, and Kashmir. ***World:*** Abyssinia, Iran, Afghanistan, India, and Nepal.

(continued)

Fig. 3.45 *Debregeasia saeneb* (Forssk) Heper and Wood

Family Name:	Urticaceae
Description:	An evergreen shrub up to 3 m tall. Stem erect, branched, and woody with dark brown bark. Leaves simple, oblanceolate or lanceolate, serrate, alternate, petiolate, with upper dark green and under silvery-whitish surface. Flowers unisexual, in globose heads arranged in panicles or racemes. Calyx four-lobed, 3–5 exserted stamens. Female flowers in smaller clusters than male. Fruit achene, yellowish-green (Fig. 3.45).
Medicinal Uses:	
Collection:	60–70 g of fresh leaves is collected by men, women, and children 14–40 years old, in summer (March–September).

(continued)

Family Name:	Urticaceae
Recipes:	60–70 g of fresh leaves is ground daily for 5–6 min; 1½ cup (375 mL) of water is added to the ground leaves. Then the mixture is filtered with a piece of cloth or a filtration pot and given to patients suffering from diarrhea, dysentery, and earaches. **For children,** 2 tsp (15–15 mL) of drug (at one time) is mixed in a cup (120 g) of curd and given 2–3 times per day for 2–3 days. **For adults,** ½ cup (125 mL) of drug (at one time) is mixed in 2 cups (500 g) of curd and is given 2–3 times per day for 3–4 days. **For earache**, 2–3 drops (3 mL) of drug (at one time) are placed in the ear 2–3 times per day for 1–2 days.
Diseases Cured:	Dysentery, diarrhea, and earache.
Ethno botanical Uses:	Leaves are used as fodder for goats and sheep. Fresh fruit is edible (sweet). The plant is used as fuelwood and for mud roof thatching.

3.4.28 Dodonaea viscosa Linn.

Family Name:	Sapindaceae
Local Name/English Name:	Sanatha, Ghuraskay, Banmandra/Switch sorrel
Flowering Period:	January–March
Status:	Common
Part Used:	Whole plant
Habit/Habitat:	A medium-sized shrub, found on exposed dry slopes in dry hard clay.
Distribution:	***Pakistan:*** Karachi, Sind, Balouchistan, Kurram, Hazara, Salt Range, Dir, and Swat. ***World:*** Found in pan-tropical areas.
Description:	An evergreen medium-sized shrub with resinous parts. Stem erect, branched, and woody with reddish-brown bark. Leaves simple, alternate, sessile, entire, bright green, shining, oblanceolate, and fleshy. Male flowers have 6–8 stamens and red anthers. Female flowers with three-angled ovary; petals are absent. Fruit winged capsule (Fig. 3.46).
Medicinal Uses:	
Collection:	½ kg of fresh leaves and 250 g of fresh fruit are collected by men, women, and children 12–40 years old, in summer (April–August).

(continued)

Fig. 3.46 *Dodonaea viscosa* Linn

Family Name:	Sapindaceae
Recipes:	**(a)** ½ kg of fresh leaves is boiled daily in 1½ L of water for 15–20 min. Patients' joints and swollen body parts are placed in the steam coming from this boiling pot, whereas the paste made from the leaves (*malum*) is applied to infected body parts and secured with a piece of cloth. The same method is used once or twice a day, especially at bedtime, for 5–6 days. **(b)** 250 g of fresh fruit is boiled in 1 L of water for 20–25 min.; when 3 cups (750 mL) of water remains, the water is filtered with a piece of cloth; given to patients suffering from intestinal worms. **For children,** 1–2 tsp (8–10 mL) of decoction (at one time) are given 2–3 times per day for 2–3 days. **For adults,** ½ cup (125 mL) of decoction (at one time) is given 2–3 times per day for 3–4 days.
Diseases Cured:	Rheumatism, swelling, and intestinal worms.

(continued)

Family Name:	Sapindaceae
Ethnobotanical Uses:	Leaves are also used as fodder for goats and sheep. The plant is used as fuelwood, in mud roof thatching, and to make brooms and walking sticks. Young elastic branches are used to make rope (*sub*).
Phytochemicals:	Acid, resin, gum, albumin, tannin, ash, alkaloid, and saponin [43].

3.4.29 Euphorbia prostrata Ait

Family Name:	Euphorbiaceae
Local Name/English Name:	Lal booti/Prostrate sandmat
Flowering Period:	April–December
Status:	Uncommon
Part Used:	Leaves
Habit/Habitat:	A prostrate herb, grows sandy waste places.
Distribution:	***Pakistan:*** Karachi, Las Bela, Mach, Hazara, Islamabad, Pabbi, Pasrur, and Sakeser. ***World:*** India and Java. Native to tropical and subtropical America.
Description:	A small prostrate herb. Stem prostrate, weak, branched. Leaves simple, opposite, sessile, entire, oval-shaped. Branches small, soft, pinkish or red with milky latex. Flowers small and reddish in color (Fig. 3.47).
Medicinal Uses:	
Collection:	125 g of fresh leaves is collected by men and women 20–40 years old, in summer (April–September), cleaned, and washed in water 2–3 times.
Recipes:	125 g of fresh leaves is ground daily for 6–7 min; 30 g of sugar and 3 cups of water are also mixed in. Then the mixture is filtered and given to patients suffering from diarrhea and dysentery. **For children,** 2–3 tsp (15–20 mL) of drug (at one time) is given 2–3 times per day for 2–3 days. **For adults,** 1 cup (250 mL) of drug (at one time) is given 2–3 times per day for 2–3 days.

(continued)

Fig. 3.47 *Euphorbia prostrata* Ait

Family Name:	Euphorbiaceae
Diseases Cured:	Diarrhea and dysentery.
Phytochemistry:	Gallic acid, corilagin, 1, 2, 3-tri-O-galloyl-D-glucose, geraniin, tellimagradin I, II, rugosin A, rugosin E, rugosin D, and rugosin G [8].

3.4.30 Ficus carica Forssk.

Family Name:	Moraceae
Local Name/English Name:	Phagwar, Inzar/Fig tree
Flowering Period:	June–December
Status:	Common
Parts Used:	Bark, fruits, and latex
Habit/Habitat:	A medium-sized tree, found in cultivated places.
Distribution:	***Pakistan:*** Gilgit, Chitral, Dir, Swat, Mardan, Hazara, Peshawar, Rawalpindi, Sargodha, D. I. Khan, and Baluchistan. ***World:*** India, Pakistan, Afghanistan; Russia, Iran, Middle East, N. Africa, and Europe.

(continued)

Fig. 3.48 *Ficus carica* Forssk

Family Name:	Moraceae
Description:	A small deciduous tree with several spreading branches. Bark smooth, gray or dull white. Leaves glabrous to tomentose with lamina variable in shape and size, broadly ovate to nearly orbicular, undivided or obscurely palmatifid to mostly palmatipartite. Hypanthodia axillary solitary or paired, yellowish to brownish violet (Fig. 3.48).
Medicinal Uses:	
Collection:	Fresh latex and fruits are collected by young children or women. Latex can be collected at any time, whereas fruits are collected in May–July.
Recipes:	**(a)** Latex is obtained from the leaves or young branches and placed on the patient's body part where a prickle has hidden; the prickle is easily drawn out from the outer covering of the body. **(b)** Fresh fruits are eaten by both children and adults to cure constipatation.
Diseases Cured:	To remove prickles and cure constipatation.
Ethnonbotanical Uses:	Leaves are used as fodder for cattle. Wood is used as fuel, in shelters, and as tool handles.
Phytochemistry:	Arabinose, β-amyrins, β-carotines, glycosides, β-setosterols, xanthotoxol, psoralen, bergapten, umbelliferone, campesterol, stigmasterol, fucosterol, and fatty acids [36, 44–47].

Fig. 3.49 *Ficus varigata* Wall. ex Roxb

3.4.31 Ficus virgata Wall. ex Roxb.

Family Name:	Moraceae
Local Name/English Name:	Phagwara, Rhumbul/Fig
Flowering Period:	April–November
Status:	Common
Part Used:	Whole plant
Habit/Habitat:	A deciduous tree, mostly found in waste places along with cultivated fields in clay.
Distribution:	***Pakistan:*** Found in all four provinces. ***World:*** East Africa, Arabia, Peninsula, Iran, Afghanistan, India, and Nepal.
Description:	A medium-sized deciduous tree up to 10 m tall. Stem erect, smooth, branched, and woody with gray bark. Leaves simple, alternate, long, and ovate or palmately divided, three- to five-nerved, petiolate, toothed, pubescent on the upper surface and green. Fruit rounded, green axillary solitary or paired, pale yellowish to purple, with many seeds (Fig. 3.49).

(continued)

Family Name:	Moraceae
Medicinal Uses:	
Collection:	1 kg of ripe fruit and 2–3 tsp of fresh juice are collected by men and women 20–40 years old. Fruit is collected in summer (June–September), dried in sunlight for 4–5 days, and then stored in cloth or a glass/plastic bottle for further use. The milky juice is collected in summer (April–September), by plucking leaves or young branches, and used fresh.
Recipes:	**(a)** 125 g of dried fruit of *Ficus variegata,* 100 g of dried seeds of *Amaranthus viridis,* and 50 g of sugar are ground together for 8–10 min. Powdered drug is stored in a glass/plastic bottle and given to patients suffering from vision disorders. **For children,** not used. **For adults,** 2 tsp (10–12 g) of powdered drug (at one time) is given twice daily (morning–evening) for 14–15 days. **(b)** 2 tsp (15–15 mL) of fresh juice is mixed in 2 tsp (10–15 mL) of milk; 5–6 g of the resulting paste (*malum*) is applied on pimples, skin lesions, and other infectious body parts 2–3 times per day for 3–4 days. The drug is white.
Diseases Cured:	Vision disorderse, skin infections, pimples, and lesions.
Ethnobotanical Uses:	Leaves are used as fodder for cattle, goats, and sheep. Young leaves are cooked as spinach (*sag*) in diluted milk (*lasi*). Ripe fruit is edible, and unripe fruit is cooked as food. Leaves are also used to clean milk pods. Wood is used to make agricultural implements and tool handles, as fuelwood, and in thatching; the plant is also used as a shade tree.
Phytochemicals:	Resin, albumin, cerin, sugar, malic acid, renin, B-amyrin, ascorbic acid, and adrenaline [48].

Fig. 3.50 *Fumaria indica* (Hausskan) Pugsley

3.4.32 Fumaria indica (Hausskan) Pugsley

Family Name:	Fumariaceae
Local Name/English Name:	Papra, Pit papra, Shatara/Fumitory
Flowering Period:	March–May
Status:	Common
Part Used:	Whole plant
Habit/Habitat:	An annual herb, found as weed in wheat fields and in waste places in clay loam.
Distribution:	***Pakistan:*** Sind, Balouchistan, Hazara, Kurram Agency, Swat, and Kashmir. ***World:*** West Asia and also introduced everywhere.
Description:	A small annual, diffuse, and much branched herb up to 1 ft in height. Stem erect, branched, herbaceous, soft and green, with watery juice. Leaves simple, green, narrow, flat, much divided, segmented, and grayish-green. Flowers small, pale pink, pedicellated, born on short racemes up to 2 cm long. Fruit rounded, globose, one-seeded nutlets (Fig. 3.50).

(continued)

Family Name:	Fumariaceae
Medicinal Uses:	
Collection:	1 kg of fresh plant material is collected by men and women 20–40 years old, in summer (March–June). Then it is cleaned, washed in water 2–3 times, and dried in shade for 6–7 days.
Recipes:	**(a)** 250 g of dried plant material is ground for 6–7 min. This powder is then stored in a glass or plastic bottle and given to patients suffering from skin infections, pimples, lesions, and constipation and to purify blood. **For children,** ½ tsp (5–6 g) of drug (at one time) is given with 1 cup (250 mL) of water twice daily (morning–evening) for 7–8 days. **For adults,** 2 tsp (12–15 g) of drug (at one time) is given with 1 cup (250 mL) of water for 10–15 days. **(b)** 250 g of dried plant material is boiled in 3–4 cups (750–1000 mL) of water for 10–15 min; 4–5 tsp (50 g) of sugar is mixed in. When 2 cups (500 mL) of water is left, it is strained with a piece of cloth, and the decoction is stored in a glass bottle; given to patients suffering from fever, constipation, or skin infections and to purify blood. **For children,** 1–2 tsp (8–10 mL) of decoction (at one time) is given 2–3 times per day for 4 days. **For adults,** 2 tsp (12–15 mL) of decoction (at one time) is given 2–3 times per day for 6–7 days.
Diseases cured:	Fever, constipation, pimples, lesions, skin infections and to purify blood.
Phytochemicals:	Fumaric acid, fumarine (alkaloid) and a crystalline organic base, isoquinoline; paprafumine, paprarine, papraline, and secophthalideisoquinoline [30, 49].

3.4.33 Grewia tenax Drum. ex Burret

Family Name:	Ulmaceae
Local Name/English Name:	Dhaman, Pastaoni, Phalsa/White Cross berry
Flowering Period:	April–August
Status:	Common

(continued)

Fig. 3.51 *Grewia tenex* Drum. ex Burret

Family Name:	Ulmaceae
Part Used:	Whole plant
Habit/Habitat:	A medium-sized tree, grows in waste places in hard soil.
Distribution:	***Pakistan:*** Balouchistan, Hazara, Cherat, Salt Range, Skaser, Nurpur, Rawalpindi district, Murree, Jhelum Valley, and Kashmir. ***World:*** Found in tropical Africa, India, and Nepal.
Description:	A medium-sized tree up to 8 m tall. Stem erect, branched, hairy, and woody with yellowish-gray bark. Leaves simple, alternate, ovate or broadly so, petiolate, dentate, and green. Flowers creamy white in axillary cymes of up to six. Fruit drupe two- to four-lobed (Fig. 3.51).
Medicinal Uses:	
Collection:	1 kg of fresh bark is collected by men and women 20–40 years old, in winter (September–February). Then the bark is cut into small 2–3-in. pieces.

(continued)

Family Name:	Ulmaceae
Recipes:	1 kg of fresh bark is boiled in 3 L of water for 1–2 h. When ½ liter of water remains, it is filtered with a cloth and stored in a glass bottle; given to patients suffering from constipation and stomach inflammation. **For children,** 1–2 tsp (10–12 mL) of decoction (at one time) is given 2–3 times per day for 1–2 days. **For adults,** ½ cup (125 mL) of decoction (at one time) is given twice daily (morning–evening) for 5–6 days.
Diseases Cured:	Constipation and stomach inflammation.
Ethnobotanical Uses:	Leaves are used as fodder for cattle, goats, and sheep. Bark is used to wash hair (especially women). Wood is used to make tool handles and as firewood when dry. The young elastic branches are placed in a stream of water in the winter (December–February); in the summer (April–July), their bark is peeled off. The fibers collected from sticks are used in making ropes.
Phytochemicals:	Proteins and phenols [50].

3.4.34 Hedera helix Alin Auct

Family Name:	Araliaceae
Local Name/English Name:	Harbumbal, Parvata, Banda/Bind wood, ivy
Flowering Period:	July–August
Status:	Common
Part Used:	Leaves
Habit/Habitat:	Wild shrub, mostly found as climbing shrub on trees such as *Pyrus* and *Punica* species and on moist, cold, and shady cliffs.
Distribution:	***Pakistan:*** Swat, Hazara, Margalla Hills, Murree Hills, and Kashmir. ***World:*** Afghanistan, India, Bhutan, Southwest China, and Burma.

(continued)

Fig. 3.52 *Hedra helix* Alin Auct

Family Name:	Araliaceae
Description:	A climbing shrub, extensive climber on trees. Stem woody, branched; climber bears adventitious roots that arise from stem. Leaves are simple, petiolate, dark green, shining above and light green under surface, with three to four triangular lobes or entire. Flowers yellowish-green, 5-merous. Inflorescence panicle of umbels; fruit is black or orange, pea-like shape and size; berry globose to subglobose with 2–3 seeds (Fig. 3.52).
Medicinal Uses:	
Collection:	4–5 fresh leaves are collected by men or women 30–50 years old, in any season when required.
Recipes:	There is no special method of drug preparation. Just a few leaves (at one time) are chewed 3–4 times per day during a blood sugar problem to reduce a high level. This drug is only used by adults 30–50 years old and is not given to children.
Disease Cured:	Diabetes.

(continued)

Family Name:	Araliaceae
Ethnobotanical Uses:	The leaves are used as fodder for goats and sheep.
Phytochemicals:	Arsenic oxide, saponin, α-hedrin and glucoside helixin, arsenic oxide, saponin, hedrin. A glucoside (helexin), saponin, and glycosides [30, 51].

3.4.35 Indigofera gerardiana Wall.

Family Name:	Papilionaceae
Local Name/ English Name:	Kainthi, Ghwareja/Cool indigo
Flowering Period:	May–June
Status:	Common
Part Used:	Whole plant
Habit/Habitat:	A small shrub growing in humus soil.
Distribution:	***Pakistan:*** Hazara, Swat, Murree, Kashmir, Dir, Chitral, and Gilliyat. ***World:*** India, Nepal, Bhutan, China, Afghanistan.
Description:	A small deciduous shrub with whitish branches. Leaves imparipinnately compound, leaflets long, elliptic, obovate or oblanceolate, obtuse, mucronate, pubescent on both sides. Inflorescence a raceme, nearly sessile, pale red. Fruit long, straight, cylindric glabrous, 10–12-seeded (Fig. 3.53) *(Indigeofera).*
Medicinal Uses:	
Collection:	200 g of roots and bark is collected by men 25–40 years old, in winter (October–December), and cleaned with water.
Recipes:	**(a)** The roots are cleaned, dried, and boiled in water for 40–50 min, then cooled down and filtered. This filtrate is used against abdominal pain. **For children,** ½ cup (125 mL) of decoctio (at one time) is given once or twice daily for 4–5 days. **For adults,** 1–2 cups (250–500 mL) of decoction (at one time) is given twice daily (morning–evening) for 4–5 days. **(b)** Patients chew the bark of the roots to relieve abdominal pain. **(c)** The bark is boiled in milk and a bandage is formed. This bandage is used externally to treat cracked and broken bones.

(continued)

Fig. 3.53 *Indigofera garadiana* Wall

Family Name:	Papilionaceae
Diseases Cured:	Abdominal pain, cracked and broken parts of body.
Ethnobotanical Uses:	Used in making roofs of huts and houses, and as fodder and fuel. Branches are used for sweeping, packing, and basket making.
Phytochemistry:	Lactone, linifolin, and a wax [52].

3.4.36 Ipomoea nil (Linn.) Roth

Family Name:	Convolvulaceae
Local Name/English Name:	Airla, Airl, Shine gulay/Morning glory
Flowering Period:	July–October
Status:	Common
Part Used:	Whole plant
Habit/Habitat:	A climbing, twining herb, found as weed in maize fields and in wet places in clay loam.

(continued)

Fig. 3.54 *Ipomoea nil* (Linn.) Roth

Family Name:	Convolvulaceae
Distribution:	***Pakistan:*** Hazara, Swat, Murree, Salt Range, Kashmir, Mirpur, and Jhelum. ***World:*** India, Africa, and Malaysia to North Australia.
Description:	A twining herb up to 3 m in height. Stem twining, hairy, branched, herbaceous, yellowish-green. Leaves simple, alternate, hairy, and three-lobed. Flowers large, funnel-shaped, blue and tinged smooth, sub-globose, with 3–5 black seeds (Fig. 3.54).
Medicinal Uses:	
Collection:	250 g of dried seeds is collected by men 25–40 years old, in winter (October–December), then dried in sunlight for 1–2 days, and stored in cloth or a glass or plastic bottle for further sue.

(continued)

Family Name:	Convolvulaceae
Recipes:	250 g of dried seeds is ground for 8–10 min. This powder is then stored in a glass or plastic bottle and given to patients suffering from constipation, stomach disorders, and intestinal worms. **For children,** 1 tsp (5–6 g) of powder drug (at one time) is given with 1 cup (250 mL) of water once or twice daily for 4–5 days. **For adults,** 1–2 tsp (10–12 g) of powdered drug (at one time) is given with 1 cup (250 mL) of water twice daily (morning–evening) for 10–15 days.
Diseases Cured:	Stomach disorder, constipation, and intestinal worms.
Ethnobotanical Uses:	The fresh plant is used as fodder for domestic animals.
Phytochemistry:	Protein, calcium, phosphorus, and resin [6].

3.4.37 *Justicia adhatoda (Linn.) Nees*

Family Name:	Acanthaceae
Local Name/ English Name:	Bhekkar, Arusha, Bhekkar/Malabar nut
Flowering Period:	December–April
Status:	Common
Part Used:	Whole plant
Habit/Habitat:	A perennial branched shrub, mostly grows in shady and waste places in clay.
Distribution:	***Pakistan:*** Hazara, Chitral, Swat, N. Waziristan, Kurram, Kohat, Khyber, Karachi, Sind, and Lower Balouchistan. ***World:*** West Nepal, India, and Indochina.
Description:	An evergreen gregarious shrub, up to 3–6 m tall. Stem erect, gregarious, branched, woody, and gray. Leaves simple, elliptic–lanceolate, apposite, petiolate, entire, yellow-green. Flowers small, white, nectarous in terminal or axillary spikes. Bracts are elliptic oblong. Fruit is capsule, club-like (Fig. 3.55).
Medicinal Uses:	
Collection:	½ kg each of roots, branches, leaves, and flowers is collected by men, women, and children 10–40 years old, during winter (August–February). All these parts are used in both fresh and dried forms. They are dried in shade for 4–8 days and then stored in cloth bags for further use.

(continued)

Fig. 3.55 *Justicia adhatoda* (Linn.) Nees

Family Name:	Acanthaceae
Recipes:	**(a)** 250 g of dried leaves of *Justacia adhatoda,* 150 g of *Trachyspermum ammi* (Ajwain), 100 g of *Foeniculum vulgare* (Sounf), 80–90 g of *Terminalia belerica* (Harir), 50 g of *Zingiber officinales* (Adrak), and 70 g of "Black piper" are mixed and ground for 15–20 min. This powder is stored in a glass or plastic bottle and given to patients suffering from cough, asthma, tuberculosis, stomach problems, and phlegm. **For children,** 3–6 g of powdered drug (at one time) is given with 1 cup (250 mL) of water, 2–3 times per day, for 8–10 days. **For adults,** 10–15 g of powdered drug (at one time) is given with 1 cup (250 mL) of water, 2–3 times per day, for 10–15 days.

(continued)

Family Name:	Acanthaceae
	(b) ½ kg of fresh roots is cleaned and washed 2–3 times with water and cut into small 1–2-in. pieces; then they are boiled in 5–6 cups (1,500 mL) of water for 20–25 min. When 2 cups (500 mL) of water is left, it is filtered with a piece of cloth. Next, 5–6 g of "alum" (*patkhri*) is added and again boiled for 4–5 min to remove dirty froth. Then it is chilled and stored in a glass bottle; given to patients suffering from cough, asthma, diabetes, gas trouble, and blood diseases. **For children,** 1–2 tsp (6–12 mL) of this decoction is mixed with 1 cup (250 mL) of water (at one time) and given once or twice daily for 5–7 days. **For adults,** 3–4 tsp (15–25 mL) of this decoction is mixed with 1 cup (250 mL) of water (at one time) and given 2–3 times per day for 10–12 days. **(c)** 300 g of fresh roots is cleaned and washed with water (2–3 times), cut into small pieces (1–2 in.), and then boiled in 4 cups (1 L) of water for 20–25 min. When 2–3 cups (500–750 mL) of water remains, it is filtered with a cloth. This decoction is stored in a glass bottle and given to patients suffering from jaundice and stomach problems. **For children,** 2–3 tsp (12–15 mL) of decoction is given daily before breakfast for 10–12 days. **For adults,** 4–5 tsp (20–30 mL) of decoction is given daily before breakfast for 15–16 days. **(d)** 200 g of fresh flowers and 200 g of sugar are mixed by hand for 5–10 min. Next, the mixture is put in a mud pot and placed in sunlight for 8–10 days. This *Gullkand* (a sweet preserve of rose petals) is given to patients suffering from cough, asthma, and chest pain. **For children,** 2–3 g of gullkand (at one time) is given 2–3 times per day for 8–10 days. **For adults,** 6–8 g of gullkand (at one time) is given 2–3 times per day for 10–15 days. **(e)** 60–70 g of root, leaves, branches, or flowers, which are already dried, are burned on an iron plate; the ashes are stored in a plastic bottle. Used for teeth and gum pains; 2–3 g of ash powder (at one time) is rubbed on teeth 3 times per day, after meals, for 15–20 days.

(continued)

Family Name:	Acanthaceae
	(f) ½ kg of fresh leaves is ground daily for 4–5 min; 10–15 g (2 tsp) of salt is mixed in. This ground material is given to cattle suffering from diarrhea, dysentery, gas trouble, and other stomach problems. A dose of 250 g of drug is given twice daily (morning–evening) for 2–3 days.
Diseases Cured:	Cough, asthma, bronchitis, stomach inflammation, dysentery, diarrhea, phlegm, jaundice, diabetes, mouth sores, toothaches, and tuberculosis; and to purify blood.
Ethnobotanical Uses:	It is used as firewood when dried. Leaves are grazed by goats and sheep.
Phytochemicals:	Essential oil, fats, resin, vasicine, vasicol, vasicisone, peganine, adhatodic acid, sugar, gum, coloring matter, and salts [53–55].

3.4.38 Mallotus philippensis (Lam.) Muell. Arg.

Family Name:	Euphorbiaceae
Local Name/English Name:	Kamila, Kamala, Kumile/Indian kamala
Flowering Period:	September–November
Status:	Uncommon
Part Used:	Whole plant
Habit/Habitat:	A small tree, found mostly in forest areas in waste places in dry hard soil.
Distribution:	***Pakistan:*** Sind, Sawat, Kashmir, and Lower Hazara. ***World:*** India, Sri Lanka, Indochina, Malaysia, Australia, and Polynesia.
Description:	A small tree up to 5 m tall. Stem erect, branched, and woody, with gray or pale brown rough bark. Leaves simple, alternate, petiolate, entire, and longer than broad, lanceolate, with upper surface dark green and red glandular beneath. Flowers numerous, small, cream color, in erect short clusters of spikes. Fruit subglobose, three lobes covered with dense reddish powdery covering. Seeds are black and smooth (Fig. 3.56).

(continued)

Figure 3.56 Mallotus philippensis (Lam) Muell. Arg

Fig. 3.56 *Mallotus philippensis* (Lam) Muell. Arg

Family Name:	Euphorbiaceae
Medicinal Uses:	
Collection:	½ kg of dried fruit is collected by men 25–40 years old, in winter (December–February). The fruits' seed coats are removed; the seeds are dried in sunlight for 1–2 days and stored in cloth or a glass jar for further use.
Recipes:	250 g of dried seeds is ground for 10–15 min; this powder is stored in a glass or plastic bottle and given to patients suffering from constipation and intestinal worms. **For children,** ½ tsp (3–4 g) of powdered drug (at one time) is mixed with ½ cup (125 g) of curd and given 2–3 times per day for 1–2 days. **For adults,** 1 tsp (10–12 g) of powdered drug (at one time) is mixed with 1 cup (250 g) of curd and given 2–3 times per day for 5–6 days.

(continued)

Family Name:	Euphorbiaceae
Diseases Cured:	Intestinal worms and constipation.
Ethnobotanical Uses:	Leaves are used as fodder for cattle, goats, and sheep. Wood used in thatching and tool handles and as fuel.
Phytochemistry:	Cardeniotids, corotonigenin, l-rhamanoside, and coroglaucigenin rhamnoside, bergenine, 4-hydroxyrottlerine, 3, 4-dihydroxy rottlerine, and 4-phloroglucinol derived phorbic acid [1].

3.4.39 *Melia azedarach Linn.*

Family Name:	Meliaceae
Local Name/English Name:	Drek, Bakayan, Dhek/Common bead tree
Flowering Period:	March–April
Status:	Common
Part Used:	Whole plant
Habit/Habitat:	A common tree, found as cultivated as well as wild plant in moist ravines in waste places and near houses.
Distribution:	***Pakistan:*** Found as a favorite garden and roadside tree in all four provinces. ***World:*** Found as a wild plant in Himalaya, as a cultivated plant in Iran, India, China, Burma, and Turkey.
Description:	A medium-sized tree up to 10 m tall. Stem erect, branched, and woody, with dark gray bark. Leaves compound, 7–10 leaflets, green, toothed, opposite, short, and petiole. Flowers violet, small, numerous in clusters. Fruit globose, yellow, and single-seeded (Fig. 3.57).
Medicinal Uses:	
Collection:	60–70 g of fresh leaves and 250 g of dried fruit are collected by men 20–40 years old. Leaves are collected in spring (March–May) and fruit is collected in summer (June–September). Fruit is also stored in cloth for further use.

(continued)

Fig. 3.57 *Melia azedrach* Linn

Family Name:	Meliaceae
Recipes:	**(a)** 30 g of fresh leaves is boiled daily for 7–8 days in 3 cups (750 mL) of water for 15–20 min. When 1 cup (250 mL) of water is left, it is strained with a cloth or filtration pot. This decoction is given to patients suffering from eye diseases (eyesore) and malarial fever. **For children,** 1 tsp (5 mL) of decoction (at one time) is mixed in 1 cup of water, 5–6 g of salt is also added, and then the mixture is given to the patient 2–3 times per day for 6–7 days. **For adults,** 2–3 tsp (15–20 mL) of decoction (at one time) is mixed in 1 cup (250 mL) of water, 8–10 g of salt is also added, and the mixture is given 2–3 times per day for 8–10 days. **(b)** 75 g of dry fruit of *Melia azadarach*, 125 g dry fruit of *Phyllanthus emblica* (Amla), and 125 g of *Terminalea chebula* (Harir) are ground together for 10–15 min. Then 25–30 tablets (each 5–6 g) are made from this powder and stored in a plastic or glass bottle for further use. These tablets are given to patients suffering from piles, and to purify the blood. **For children,** not used. **For adults,** one tablet (5 g) is given twice daily (morning–evening) for 10–15 days.

(continued)

Family Name:	Meliaceae
	(c) 50 g of fresh leaves is ground daily for 5–6 min. This paste (*malum*) is used for headaches and body swelling due to injury. Ten to 15 g of the paste (at one time) is applied on the infected area 2–3 times per day for 4–5 days.
	(d) 70–80 g each of fresh leaves, twigs, and fruit is ground daily for 8–10 min. This drug is given to cattle suffering from gas trouble and indigestion. A dose of 125 g of drug is given twice daily (morning–evening) for 2–3 days.
Diseases Cured:	Malarial fever, piles, eye ache, headache, swelling, and wounds; to purify blood; and gas trouble and indigestion in cattle.
Ethnobotanical Uses:	Fresh leaves are used as fodder for cattle, goats, and sheep. Wood is used to make agricultural implements, tool handles, thatching, and light furniture, and as timber and fuelwood.
Phytochemicals:	Bakayanin, margosine, alkaloid (azedarin), resin, tannin, meliotannic acid, and benzoic acid [56].

3.4.40 *Mentha longifolia (Linn.) Huds*

Family Name:	Lamiaceae
Local Name/ Family Name:	Chitta podna, Enalay, Baburi/Horse mint
Flowering Period:	July–August
Status:	Common
Part Used:	Leaves
Habit/Habitat:	A perennial herb, grows along water courses on wet places and also cultivated near houses in pots or in small land patches.
Distribution:	***Pakistan:*** Hazara, Kashmir, Poonch, Swat, Chitral, Kurram, and Baltistan. ***World:*** Northwest Asia, Europe, India, Nepal, and South Africa.
Description:	A small, erect, rhizomatous and aromatic herb up to 1 ft in height. Stem is erect, branched, 4-angled, herbaceous, hairy, green or purple. Leaves simple, elliptic–oblong to oblanceolate, opposite, petiolate, and toothed, hairy, whitish, aromatic. Flowers in whorls on axillary or terminal spikes, small, pink or mauve to violet, numerous (Fig. 3.58).

(continued)

Fig. 3.58 *Mentha longifolia* (Linn), Huds

Family Name:	Lamiaceae
Medicinal Uses:	
Collection:	2 kg of fresh leaves is collected by men, women, and children 12–50 years old, in summer (March–August). Leaves used in both fresh and dried forms; dried in shade for 5–6 days, then stored in cloth sacks for further use.
Recipes:	**(a)** 70 g of dried leaves of *Mentha longifolia,* 30–40 g of *Trachyspermum ammi* (Ajwain), and 1–2 tsp (10–12 g) of common salt are ground together for 7–8 min. This powder is stored in a plastic bottle or paper bag and given to patients suffering from indigestion, gas trouble, and stomach disorders. **For children,** 1 tsp (10–15 g) of powdered drug (*phaki*) (at one time) is given with 1 cup (250 mL) of water 2–3 times per day for 2–3 days. **For adults,** 2 tsp (10–15 g) of powdered drug (at one time) is given with 1 cup (250 mL) of water 2–3 times per day for 2–3 days.
	(b) 70 g of dried leaves of *Mentha longifolia,* 2–3 tsp (20–25 g) of *Trachyspermum ammi* (Ajwain), and 2–3 tsp (20–25 g) of sugar are boiled in 3 cups (750 mL) of water for 8–10 min. Then it is filtered with a cloth or filtration pot and given to patients suffering from cholera, fever, cough, and vomiting. **For children,** ½ cup (125 mL) of decoction (at one time) is given 2–3 times per day for 3–4 days. **For adults,** 1 cup (250 mL) of decoction (at one time) is given 2–3 times per day for 4–5 days.

(continued)

Family Name:	Lamiaceae
	(c) 30 g of dry leaves of *Mentha longifolia,* 15 g of dried seeds of *Punica granatum* (Druna), 10–12 g of fresh scales of *Allium cepa* (Piaz), and 1 tsp (5–6 g) of common salt are ground together daily for 7–8 min and then 2–3 cups of water is added. This mixture is shaken well and then filtered with a cloth or filtration pot. This juice is given to patients suffering from cholera, indigestion, vomiting, stomach disorders, and gas trouble. **For children,** ½ cup (125 mL) of drug is given (at one time) 2–3 times per day for 1–2 days. **For adults,** 1–cup (250 mL) of drug is given (at one time) 3–4 times per day for 1–2 days.
Diseases Cured:	Stomach disorders, gas trouble, indigestion, vomiting, cholera, fever, and cough.
Ethnobotanical Uses:	Fresh and dried leaves are used in chutneys; also used as spice, stimulant, and carminative in curries.
Phytochemicals:	Oleanolic, ursolic and fatty acids, menthol, essential oil, linalool, (+, -), alcohol, carvone, pulegone, luteolin-7 glycoside, luteolin-7-glucuroside, luteolin-7-glucuronide, apigenin 7-glucuronide, aceacetin 7-rutinoside, liosmetin 7-rutlnoside, lesperetin 7-rutinoside, eriodictyol 7-rutinoside, acacetin and eriodictyol, a-pinene, myrcene, limonene, alloocymene, caryophyllene, humelene, carvone, volatile oil, thymol, resin, gum, and tannin [1, 56].

3.4.41 *Mentha royleana Linn.*

Family Name:	Lamiaceae
Local Name/English Name:	Kala podna, Venalay, Pudina/Mint
Flowering Period:	July–August
Status:	Common
Part Used:	Leaves
Habit/Habitat:	A perennial herb, grows along water courses in wet places and also cultivated near houses in pots or in small land patches.
Distribution:	***Pakistan:*** Hazara, Kashmir, Poonch, Swat, Chitral, Kurram, and Baltistan. ***World:*** Northwest Asia, Europe, India, Nepal, and South Africa.

(continued)

Fig. 3.59 *Mentha royleana* Linn

Family Name:	Lamiaceae
Description:	A small, erect, rhizomatous, and aromatic herb up to 1 ft in height. Stem erect, branched, 4-angled, herbaceous, hairy, green or purple. Leaves simple, elliptic–oblong to oblanceolate, opposite, petiolate, and toothed, hairy, dark green, and aromatic. Flowers in whorls on axillary or terminal spikes, small, whitish or of cream color, and numerous (Fig. 3.59).
Medicinal Uses:	
Collection:	2 kg of fresh leaves is collected by men, women, and children 12–50 years old, in summer (March–August). Leaves used in both fresh and dried forms; dried in shade for 5–6 days and then stored in cloth sacks for further use.
Recipes:	30 g of dried leaves of *Mentha roylena,* 15 g of dried seeds of *Punica granatum* (Druna), 10–12 g of fresh scales of *Allium cepa* (Piaz), and 1 tsp (5–6 g) of common salt are ground together daily for 7–8 min and mixed with water. This mixture is shaken well, filtered with a cloth or filtration pot, and given to patients suffering from cholera, indigestion, vomiting, stomach disorders and gas trouble. **For children**, ½ cup (125 mL) of drug is given (at one time) 2–3 times per day for 1–2 days. **For adults,** 1 cup (250 mL) of drug is given (at one time) 3–4 times per day for 1–2 days.
Diseases Cured:	Stomach disorders, gas trouble, indigestion, vomiting, and cholera.

(continued)

Fig. 3.60 *Myrisine africana* Linn

Family Name:	Lamiaceae
Ethnobotanical Uses:	Fresh and dried leaves are used in chutneys; also used as spice, stimulant, and carminative in curries and salad.
Phytochemicals:	Volatile oil, menthol, resin, tannin, and gum [36].

3.4.42 Myrsine africana Linn.

Family Name:	Myrsinaceae
Local Name/ English Name:	Khukan, Babrung, Baobirung/African boxwood"
Flowering Period:	March–May
Status:	Common
Part Used:	Whole plant
Habit/Habitat:	A common shrub in forests, found mostly in waste shady places in loamy soil.
Distribution:	***Pakistan:*** Balouchistan, Chitral, Peshawar, Swat, Hazara, and Kashmir. ***World:*** Africa, Arabia, Afghanistan, Tibet, China, Taiwan, India, and Nepal.
Description:	An evergreen shrub up to 2 m tall. Stem erect, branched, and woody, with grayish-brown bark. Leaves simple, alternate, entire, and lance-shaped, subsessile, green. Flowers small, dispersed in racemes, numerous, minute, and greenish-red. Fruit globose, drupe, dark blue to black, single-seeded (Fig. 3.60).

(continued)

Family Name:	Myrsinaceae
Medicinal Uses:	
Collection:	1 kg of ripe fruit and 25–30 g of fresh leaves are collected by men and children 12–40 years old, in summer (June–August). Fruit is dried in sunlight for 4–5 days and stored in cotton sacks or a glass or plastic bottle for further use, whereas leaves are used in fresh condition.
Recipes:	**(a)** 25 g of dried fruits is ground for 5–6 min. This powdered drug is stored in a glass or plastic bottle and given to patients suffering from intestinal worms. **For children,** 1 tsp (4–6 g) of powdered drug (at one time) is given with 1 cup (50 g) of curd or 1 cup (120 mL) of diluted milk (*lasi*) once daily, at bedtime, for 3–4 days. **For adults,** 2 tsp (8–12 g) of powdered drug (at one time) is given with 2 cups (100 g) of curd or 1 cup (250 mL) of diluted milk (*lasi*) daily, at bedtime, for 3–5 days.
	(b) 20–25 g of fresh leaves is boiled daily in 3 cups (750 mL) of water for 15–20 min; when 1 cup (250 mL) of water is left, it is strained with a piece of cloth. Given to patients suffering from urinary disorders (scanty urination), kidney pain, skin diseases (allergy and pimples), and to purify blood. **For children,** 1 tsp (5 mL) of decoction (at one time) is given twice daily (morning–evening) for 6–7 days. **For adults,** ½ cup (125 mL) of decoction (at one time) is given twice daily (morning–evening) for 10–15 days. The drug's taste is bitter and its color is light yellow.
Diseases Cured:	Intestinal worms, urinary disorder (scanty urination), kidney pain, and skin diseases (allergy and pimples); also used to purify blood.
Ethnobotanical Uses:	Leaves are used as fodder for cattle, goats, and sheep. Ripe fruit is edible; stem and branches are used in brooms, mud roof thatching, and as firewood. Fruit is also sold for cash. Ropes (*sub*) are made from young elastic stems.
Phytochemicals:	Embelin, vilangin methylene bis (2, 5-dihydroxy-4-undecyl-3, 6-benzoquinone), embolic acid, querctiol [28].

3.4.43 *Olea ferruginea Royle*

Family Name:	Oleaceae
Local Name/English Name:	Kahu, Khuwana, Kao/Olive tree
Flowering Period:	April–June
Status:	Common
Part Used:	Whole plant
Habit/Habitat:	A common evergreen tree, found as self-growing plant in waste places in dry clay loam of lower elevations.
Distribution:	***Pakistan:*** Hazara, Dir, Chitral, Swat, Murree, Salt Range, and Kashmir. ***World:*** Found in Afghanistan, Nepal, and India.
Description:	A medium-sized tree up to 15 m. Stem erect, branched, woody and hard, with dark gray bark. Leaves simple, opposite, entire, petiolate, lanceolate, upper fleshy dark green and lower yellowish-green surface. Flowers small, numerous, creamish, and in terminal or lateral cymes. Fruit drupe, oval-shaped, and black (Fig. 3.61).
Medicinal Uses:	
Collection:	50 g of fresh leaves is collected by men and children 14–30 years old, in any season when required.
Recipes:	40–50 g of fresh leaves is boiled daily in 2 cups (500 mL) of water for 10–15 min. When 1 cup (250 mL) of water is left, it is strained and given to patients suffering from toothache, mouth gums, hoarseness, and throat sore. **For children,** 2 tsp (12–15 mL) of decoction (at one time) is used 3–4 times per day for 2–3 days. **For adults,** ½ cup (125 mL) of decoction (at one time) is used 3–4 times per day for 3–4 days.
Diseases Cured:	Mouth sores, toothache, throat pain, and hoarseness.

(continued)

Fig. 3.61 *Olea ferruginea* Royle

Family Name:	Oleaceae
Ethnobotanical Uses:	Leaves are used as fodder for cattle, goats, and sheep. Wood is used for making light furniture, agricultural implements, tool handles, hoe and dagger handles, mud roof thatching, and as fuelwood and timber. Young elastic branches are used to make ropes (*sub*) and sticks; the white gummy extract is edible (sweet like sugar). Fruit is used to make beads (*tusbi*). Leaves are used in making green tea (*kawa*).
Phytochemicals:	Oil starnin, arachin, cholestrin, olein, linolein, palmatin (Lvs) fixed oil; oleic, linoleic, palmatic, stearic, and myristic acids [57].

3.4.44 Otostegia limbata (Benth.) Boiss.

Family Name:	Lamiaceae
Local Name/English Name:	Koi booi, Spina ghazai, Chotta kanda/Otostegia
Flowering Period:	April–June
Status:	Uncommon

(continued)

Fig. 3.62 *Otostegia limbata* (Benth), Boiss

Family Name:	Lamiaceae
Part Used:	Whole plant
Habit/Habitat:	A spiny shrub, mostly found in waste and dry places in dry soil.
Distribution:	***Pakistan:*** Hazara, Swat, Chitral, Kurram, Peshwar, Rawalpindi, Salt Range, Kashmir, Jhelum and Islamabad. ***World:*** endemic to Kashmir (India and Pakistan).
Description:	A small, bushy, branched, slender, and spiny shrub up to 2 ft tall. Stem erect, branched, woody, spiny, whitish or gray bark. Leaves small, oblanceolate with spiny bracts and short petiole. Flowers in axillary clusters, long, pale yellow with orange throat (Fig. 3.62).

(continued)

Family Name:	Lamiaceae
Medicinal Uses:	
Collection:	50–60 g of fresh leaves is collected by men and women 20–40 years old, in summer (April–July). Leaves used in both fresh and dried forms; dried in sunlight for 1–2 days.
Recipes:	**(a)** 50 g of fresh leaves is ground daily for 3–4 min; 3–4 tsp (50 mL) of water is also added. This mixture is filtered with a cloth and given to patients suffering from mouth sores and throat pains. **For children,** ½ tsp (2.5 mL) of drug (at one time) is given twice daily (morning–evening) for 3–4 days. **For adults,** 1 tsp (5 mL) of drug (at one time) is given 2–3 times per day for 4–5 days. **(b)** 30 g of dried leaves is ground for 3–4 min. This powder is mixed in 2–3 tsp of softened butter. Ten to 15 g of this paste is applied on wounds 3–4 times per day for 4–5 days.
Diseases Cured:	Mouth sores, throat pains, and wound healing.
Ethnobotanical Uses:	Leaves are used as fodder by goats and sheep. The whole plant is used as firewood when dry.
Phytochemicals:	Saponin, pectin, and resin [28].

3.4.45 *Oxalis corniculata Linn.*

Family Name:	Oxalidaceae
Local Name/ English Name:	Jundora, Khati booti, Threwaky/Indian sorrel
Flowering Period:	March–December
Status:	Common
Part Used:	Whole plant
Habit/Habitat:	Herb, mostly grows in shady waste places and in cultivated lands in loamy soil.
Distribution:	***Pakistan:*** Found throughout all four provinces. ***World:*** Cosmopolitan.
Description:	A prostrate herb up to 6 in. in height. Stem creeping, herbaceous, branched, and green. Leaves compound, 3–5 leaflets, petiolate, obcordate, entire, pale green. Flowers yellow, pedicellate in umbels. Fruit capsule, pubescent with brown seeds (Fig. 3.63).

(continued)

Fig. 3.63 *Oxalis corniculata* Linn

Family Name:	Oxalidaceae
Medicinal Uses:	
Collection:	125 g of fresh plant material is collected daily by men, women, and children 12–40 years old, in summer (March–September), then cleaned and washed in water 2–3 times.
Recipes:	125 g of fresh plant material is ground daily for 6–7 min; 1 cup of (250 mL) of water is also mixed. Then it is strained with a cloth and given to patients suffering from stomach disorders, liver, and intestinal/stomach inflammation. **For children,** ½ cup (125 mL) of decoction (at one time) is given once daily, in the morning, for 4–5 days. **For adults**, 1 cup (250 mL) of decoction (at one time) is given once daily, in the morning, for 10–15 days. Two to 3 tsp (15–20 mL) of fresh juice (at one time) is applied 2–3 times per day on external wounds caused by injury, for 3–4 days, to kill germs and to stop bleeding from wounds.
Diseases Cured:	Stomach disorders; intestinal, liver, and stomach inflammation; wound healing; to stop bleeding; and to kill germs.

(continued)

Family Name:	Oxalidaceae
Ethnobotanical Uses:	Young leaves are used like spinach (*sag*). It is also used as fodder for cattle, goats, and sheep.
Phytochemicals:	Vitamin C, carotene, calcium, tartaric acid, citric acid, malic acid, crystalline principal, oxalate of potash and oxalic acids [56].

3.4.46 *Paeonia emodi Wall. ex Royle*

Family Name:	Paeoniaceae
Local Name/ English Name:	Mamaikh/Himalayan peony
Flowering Period:	April–May
Status:	Common
Part Used:	Rhizomes, roots, and seeds
Habit/Habitat:	A small perennial gergarious herb, found in humus soil.
Distribution:	***Pakistan:*** Dir, Chitral, Hazara, Murree, Kaghan, and Naran. ***World:*** India, eastern Afghanistan.
Description:	A small perennial herb. Leaves biternate or ternate, glabrous; lamina pale on undersurface, decurrent; median segment deeply 3-incised, the lateral segments elliptic-lanceolate or lanceolate. Flowers solitary, axillary; three to five seeds, large, globose-ovoid, minutely rugose, brown-black, bright scarlet red when unripe (Fig. 3.64).
Medicinal Uses:	
Collection:	Tubers are collected by young childrens and women 20–40 years old, washed in water 2–3 times, and dried for 15–20 days in the shade.
Recipes:	100 g of dried tubers is ground into a fine powder and cooked in *desighee*. This paste is used to cure rheumatism, backaches, and as a tonic. **For children,** 1 tsp (5 mL) and 1 cup (250 mL) of paste (at one time) is given in the morning for 10–20 days. **For adults,** 1 cup (250 mL) of the paste (at one time) is given once daily for 10–20 days.
Diseases Cured:	Rheumatism and backache; also used as a tonic.
Phytochemicals:	Oxalic acid, tannins, paeoninol and paeonin C, oligostilbene and monoterpene galactoside, 4-hydroxybenzoic acid 3, gallic acid 4, and methyl gallate [52, 58].

Fig. 3.64 *Paeonia emodi* Wall. ex Royle

3.4.47 Papaver somniferum L.

Family Name:	Papaveraceae
Local Name/English Name:	Khash-khash, doda, apeem/Opium poppy
Flowering Period:	April–June
Status:	Rare
Parts Used:	Latex and seeds
Habit/Habitat:	A small herb, found in the margins of cultivated fields and gardens.
Distribution:	***Pakistan:*** Hazara, Swat, Dir, Malakand, Peshawar, Murree, and Rawalpindi. ***World:*** Europe and Asia; known from cultivation or used as a recreational drug in W. Pakistan.
Description:	An annual herb, apparently glabrous, glaucous, rarely branched. Leaves ovate–oblong, cordate, undulate, serrate, crenate, or dentate. Flower buds are ovoid–oblong. Flowers large, showy, white, pinkish or reddish, rarely pale violet, with or without a basal dark blotch. Capsule subglabrous smooth with a rounded base. Seeds small and white (Fig. 3.65).

(continued)

Fig. 3.65 *Papaver somniferum* L

Family Name:	Papaveraceae
Medicinal Uses:	
Collection:	Fresh fruit is collected by women or children 20–40 years old, in late summer (July–August), and dried under shade for 10–150 days.
Recipes:	10–15 g of dried fruits is boiled in tea and drunk at nighttime to cure severe cough, flu, and abdominal pain. **For children,** 2–3 tsp (15–20 mL) of decoction (at one time) is given once daily, at night, for 2–3 days. **For adults,** 1 cup (250 mL) of drug (at one time) is given once daily, at night, for 5–6 days.
Diseases Cured:	Cough, flu, and abdominal pain.
Phytochemicals:	Sap contains oxalic acids. The plant contains more than 40 different alkaloids, of which morphine (up to 20%), codeine (about 1%), narcotine (about 5%), and papaverine (about 1%) are prominent. It also contains meconic acid, albumen, mucilage, sugars, resin, and wax [52].

3.4.48 Phyla nodiflora (L.) Greene

Family Name:	Verbenaceae
Local Name/English Name:	Hifza phruii, Taka aspa/Fog grass
Flowering Period:	April–September
Status:	Uncommon
Part Used:	Leaves
Habit/Habitat:	A prostrate perennial herb, found in sandy soil along water courses.
Distribution:	***Pakistan:*** Sind, Karachi, Waziristan, Kurram, Balouchistan, Parachinar, Landi Kotal, Khyber, Kohat, Lowrer Hazara, Punjab Plain, and Kashmir. ***World:*** Tropic and subtropic areas.
Description:	A small prostrate perennial herb up to 2 ft in tall. Stem prostrate, branched, green, bears adventitious roots on nodes. Leaves simple, opposite, green, fleshy, toothed on upper half and entire below. Flowers small; numerous in ovoid spikes like heads and pale white. Corolla is 2-lipped, lower lip 3-lobed (Fig. 3.66).
Medicinal Uses:	
Collection:	60–75 g of fresh leaves is collected by men and women 20–40 years old, in summer (April–August) and then washed in water 2–3 times.
Recipes:	60–70 g of fresh leaves and 8–10 g of black pepper are ground daily for 8–10 min; 1 cup (250 mL) of water is also added. Then it is filtered with a cloth and given to patients suffering from skin diseases, mouth sores, earaches, toothaches, eye diseases (*phora*), general body inflammation, and piles; also used to purify blood. **For children,** 2–3 tsp (15–20 mL) of drug (at one time) is given once daily, in the morning, for 5–6 days. **For adults,** 1 cup (250 mL) of drug (at one time) is given once daily, before breakfast, for 10–15 days.
Diseases Cured:	Skin disease, pimples, lesions, mouth sores, piles, earaches, toothaches, body inflammation, and eye diseases (*phora*); also used to purify blood.
Phytochemicals:	Glycosides (nodiflorin A, nodiflorin B), volatile oil, resin, sugar, and potassium nitrate [16].

Fig. 3.66 *Phyla nodiflora* (L.) Greene

3.4.49 Phyllanthus emblica Linn.

Family Name:	Euphorbiaceae
Local Name/English Name:	Amla/Indian gooseberry
Flowering Period:	March–April
Status:	Uncommon
Part Used:	Whole plant
Habit/Habitat:	A medium-sized tree, grows in hard, stony and rough places
Distribution:	***Pakistan:*** Hazara, Margalla Hills, and Haripur. ***World:*** India, Sri Lanka, and east to S. China and western Malaysia.
Description:	A medium-sized tree up 8 m tall. Stem erect, branched, and woody with light gray bark. Leaves simple but appear compound, numerous, light green. Flowers greenish-yellow. Fruit globose fleshy, pale yellow, with brownish-yellow seeds (Fig. 3.67).
Medicinal Uses:	
Collection:	5 kg of fresh fruit and 3 kg of bark are collected by men 18–40 years old. Fruit is collected in winter (August–November) and bark is collected in summer (March–June). First, both fruit and bark are cut with a knife into small pieces. Then they are dried in sunlight for 6–7 days and stored in cloth sacks or small baskets for further use.

(continued)

Fig. 3.67 *Phyllanthus emblica* Linn

Family Name:	Euphorbiaceae
Recipes:	**(a)** 10–15 g of dried fruit and 10–15 g of dried bark are soaked daily at night in 1 cup (250 m) of water. The next morning 1–2 tsp (10–12 g) of sugar and 2 tsp (10–12 g) of seed powder of *Punica granatum* is also mixed in; the whole mixture is shaken well and then filtered with a piece of cloth. This filtrate (*sharbat*) is given to patients suffering from jaundice, internal body inflammation, dysentery, and indigestion. **For children,** 1 cup (250 mL) of drug (at one time) is given daily, before breakfast, for 8–10 days. **For adults,** 1 cup (250 mL) of drug (at one time) is given once daily, before breakfast, for 15–20 days. **(b)** 20–30 g of dried fruit daily is boiled in 1 cup (250 mL) of water for 10–15 min, 1 tsp (6–7 g) of common salt is also mixed. This syrup is given with bread 1–2 times per day.

(continued)

Family Name:	Euphorbiaceae
Diseases Cured:	Jaundice, dysentery, fever, internal body inflammation, and indigestion.
Ethnomedicinal Uses:	Leaves used as fodder by goats and sheep. Fruit edible, eaten with salt by young children. Wood is used as fuel when dry. Fruit is also sold to make money.
Phytochemicals:	Flavonoids, kaempferol-3-O-alpha-L-(6″-methyl)-rhamnopyranoside, and kaempferol-3-O-alpha-L-(6″-ethyl)-rhamnopyranoside [59].

3.4.50 Pimpinella diversifolia (Wall.) D.C

Family Name:	Umbelliferae
Local Name/ English Name:	Tarpakhi, Sonf/Anise
Flowering Period:	August–September
Status:	Common
Part Used:	Whole plant
Habit/Habitat:	A small herb, mostly grows in waste places within grass in fertile loam.
Distribution:	***Pakistan:*** Swat, Kurram, Hazara, Chitral, Gilgit, Murree Hills, and Kashmir. ***World:*** Found throughout Himalayas, China, India, Nepal, and Burma.
Description:	A small herb up to 1 m tall. Stem erect, branched, herbaceous, and green. Leaves young undivided, simple, with serrate margin; older are more finally cut and pinnate. Flowers yellowish-white, small, numerous in umbal. Fruit small and oval-shaped (Fig. 3.68).
Medicinal Uses:	
Collection:	1 kg of fresh plant material is collected in late summer (August–October) by men and women 20–40 years old, cleaned, dried in shade for 6–7 days, and then stored in cotton sacks.

(continued)

Fig. 3.68 *Piminella diversifolia* (Wall) D.C

Family Name:	Umbelliferae
Recipes:	125 g of dried plant material *Pimpinella diversifolia* and 2–3 tsp (15–20 g) common salt are ground together and stored in a glass or plastic bottle; given to patients suffering from indigestion, abdominal swelling, leucorrhoea, gas trouble, and stomach disorders. **For children,** 1 tsp (4–5 g) of powdered drug (at one time) is given with 1 cup (250 mL) of water twice daily(morning–evening) for 8–10 days. **For adults,** 2–3 tsp (10–15 g) of powdered drug (at one time) is given with 1 cup (250 mL) water twice daily (morning before breakfast, evening at bedtime) for 8–10 days.
Diseases Cured:	Gas trouble, indigestion, abdominal swelling, leucorrhoea, and stomach disorders.
Ethnobotanical Uses:	Fresh leaves and twigs are used in chutneys and curries to improve taste and as a carminative agent.
Phytochemicals:	Essential oil, santene, a-pinene, monoterpene, myrcene, sesquiterpene, geranyl acetate, carvon, pulegone, a-terpinol, terpinolic acid, coumarins, ammirin, and oxypeucedanin [55].

Fig. 3.69 *Pistacia chinensis* Bge

3.4.51 *Pistacia chinensis Bge.*

Family Name:	Anacardiaceae
Local Name/ English Name:	Kangar, Kakra singi/Pistachio galls
Flowering Period:	March–April
Status:	Common
Part Used:	Whole plant
Habit/Habitat:	Dioecious tree, mostly found in open waste places and alongside cultivated fields in clay.
Distribution:	***Pakistan:*** Kurram Valley, Peshawar, Salt Range, Murree, Rawalpindi, Chitral, Swat, Gilgit, and Hazara. ***World:*** Afghanistan, India, and western Nepal.
Description:	A medium-sized dioecious tree up to 20 m tall. Stem is erect, branched, woody, and hard, with rough dark gray bark. Leaves compound, leaflets 7–9, lanceolate; serrate, petiolate, green with prominent nerves. The newly emerging leaves are bright red, making an attractive show in spring. Flowers unisexual, small, and reddish, in clusters. Male panicle drooping and female erect and bracteate; stamens are 3–6; styles are 3. Fruit one-seeded, drupe, grayish-brown (Fig. 3.69).

(continued)

Family Name:	Anacardiaceae
Medicinal Uses:	
Collection:	200 g of leaf galls and 300 g of fruit are collected by men 20–35 years old in summer (July–November). Galls are broken and their internal net is removed. Both galls and fruit are dried in sunlight for 4–5 days and then stored in a piece of cloth for further use.
Recipes:	**(a)** 150 g of dried leaf galls is placed on a hot iron plate, covered with another iron or silver pot, and burned for 5–10 min. This ash is then mixed with 500 g of honey. **Or** 150 g of dried galls is ground for 3–4 min. This powder is then mixed with ½ kg of honey. This drug is given to patients suffering from cough, asthma, diarrhea, and phlegm. **For children,** 1–2 tsp (10–12 g) of drug is given daily at bedtime for 10–12 days. **For adults,** 2–4 tsp (20–25 g) of drug is given daily at bedtime for 20–25 days. **(b)** 250 g of dried fruit of *Pistacia chinensis* and 250 g of previously fried grains of *Triticum aestivum* are mixed and ground together for 10–15 min; 200 g of sugar is also mixed in. This powdered drug is stored in a plastic or glass pot and given to patients suffering from gleets (*jiryan*). **For children,** not used. **For adults**, 2–3 tsp (10–15 g) of powdered drug (at one time) is given with 1 cup (250 mL) of milk or water twice a day (morning–evening) for 20–25 days.
Diseases Cured:	Cough, asthma, diarrhea, phlegm (*balghum*), and gleets (*jiryan*).
Ethnobotanical Uses:	Leaves are used as fodder for goats and sheep. Wood yields timber and is used to make furniture. It is also used as firewood and for thatching.
Phytochemicals:	Essential oil contains pinene, camphene, di-limonene, terpineol, aromadendren, caprylic acid, 2-crystalline acids, gallic acid, m-digallic acid, quercetin, 6-0-galloyl arbutin-quercitrin, and quercetin-3-0(6″-galloyl)-beta-D-glucosides [30, 60].

Fig. 3.70 *Plantago lanceolata* L

3.4.52 Plantago lanceolata L.

Family Name:	Plantaginaceae
Local Name/ English Name:	Bhatti/Narrow leaf plantain
Flowering Period:	June–September
Status:	Common
Parts Used:	Leaves, fruit, and seeds
Habit/Habitat:	A small annual herb, found in moist loamy soil along with nallah and alongside streams.
Distribution:	***Pakistan:*** Chitral, Swat, Dir, Hazara, Kashmir, Bannu, Kurram Agency, Peshawar, Sargodha, Murree Hills, Makran, D. I. Khan, and Balushistan. ***World:*** Europe, N. Africa, S. Asia to the mountains of Tien–Shan and Pamir; introduced all over the world.
Description:	A small perennial herb, up to 30 cm tall. Leaves membranous, narrow lanceolate to narrow elliptic, glabrous acute, at the base narrowed into a narrow petiole, nerves 5. Spikes are dense, conic-cylindrical, and subglobose to globose. Bracts broad ovate, narrowly caudate. Sepals long, glabrous or at the margin. Corolla lobes are narrowing ovate to ovate. Seeds are in pairs and smooth (Fig. 3.70).

(continued)

Family Name:	Plantaginaceae
Medicinal Uses:	
Collection:	100 g of fresh leaves and seeds are collected by women 25–40 years and children 15–20 years old, in spring and summer; dried in shade for 2–5 days and stored in cloth or a plastic/glass bottle for further use.
Recipes:	**(a)** Dried leaves are crushed into a fine powder. This powder is kept on wounds to stop bleeding. **(b)** 50–100 g of seed husks are soaked in water overnight. Then they are mixed with sugar; this extract is given to patients early in the morning to cure dysentery and diarrhea. **For children,** 1 cup (250 mL) of extract is given 2–3 times per day for 4–5 days. **For adults,** 2–3 cups (500 to 750 mL) of extract is given three times per day for 5–10 days.
Diseases Cured:	Dysentery and diarrhea; used to stop bleeding.
Phytochemicals:	Glucoside aucubin, resin, waxes, and large quantity of mucilage [52].

3.4.53 Plantago major Linn.

Family Name:	Plantaginaceae
Local Name/ English Name:	Bhatti/Nipple grass
Flowering Period:	May–October
Status:	Common
Parts Used:	Leaves and seeds
Habit/Habitat:	Small herb, found on moist soil along nallahs and stream sides.
Distribution:	***Pakistan:*** Swat, Hazara, Murree, Kaghan, Balouchistan, and Kashmir. ***World:*** Found throughout Europe, northern and central Asia.
Description:	A small herb. Stem is underground rhizome. Leaves simple, long, green, entire, petiolate, and oval-shaped in basel rosette. Flowers yellow, tube-shaped. Fruit capsule, opening at top with 7–8 black seeds (Fig. 3.71).
Medicinal Uses:	
Collection:	250 g of ripe seeds is collected by men and women 20–40 years old, in late summer (September–November). They are dried in sunlight for 2–5 days and stored in cloth or a plastic/glass bottle for further use.

(continued)

Fig. 3.71 *Plantago major* Linn

Family Name:	Plantaginaceae
Recipes:	250 g of dried seeds of *Plantago major* and 200 g of dried leaves of *Mentha arvensis* are ground together for 6–7 min. This powder is stored in a glass or plastic bottle and given to patients suffering from cough, asthma, dysentery, and phlegm. **For children,** 1 tsp (5–6 g) of powdered drug (at one time) is boiled in 1 cup (250 mL) of water for 5–6 min and then given twice daily for 7–8 days. **For adults,** 2 tsp (12–15 g) of powdered drug (at one time) is boiled in 1 cup (250 mL) of water for 10–15 min, and then ½ cup of "Viola syrup" is mixed in. This syrup is given to patients twice daily (morning–evening) for 15–20 days.
Diseases Cured:	Asthma, cough, dysentery, fever, and phlegm.
Ethnobotanical Uses:	Young leaves are used like spinach (*sag*) and are also used as fodder by goats, sheep, and cattle.
Phytochemicals:	Flavonoids, reducing sugar, tanning, flavonoids, alkaloids, essential oil, isoquercetrin, auculin, pectin, D-galactose, L-arabinose, L-rhamnose, chlorophyll, resin, wax, albumen, pectin, sugar, mucilage, glycoside aucubin, glycoside, and saponin [1, 61].

Fig. 3.72 *Partulaca oleracea* L

3.4.54 Portulaca oleracea L.

Family Name:	Portulacaceae
Local Name/ English Name:	Warkharay/Purslane, pusley
Flowering Period:	April–August
Status:	Common
Part Used:	Vegetative portion
Habit/Habitat:	A small herb, grows wild in maize fields.
Distribution:	***Pakistan:*** Dir, Chitral, Hazara, Kashmir, Sialkot, Lahore, Makran, Karachi, Dadu, Tharparkar, and Quetta. ***World:*** Europe, India, Ascension Island, America.
Description:	Annual or perennial herb, prostrate or erect. Stem succulent, glabrous, green or purplish-green. Leaves alternate or subopposite, closely crowded below the flowers. Inflorescence usually in the forks of branches, cymose, with clusters of three to six flowers. Flowers sessile, yellow (Fig. 3.72).
Medicinal Uses:	
Collection:	250 g of fresh plant material is collected by women and men 20–40 years old, in summer (May–July).

(continued)

Family Name:	Portulacaceae
Recipes:	**(a)** The fresh leaves and branches are ground and the juice is extracted, which is green. This juice is given to relieve abdominal pain. **For children,** 1/2 cup (125 mL) of juice is given 2–3 times per days for 4–5 days. **For adults,** 1–2 cups (250–500 mL) of juice is used thrice a day for 8–10 days. **(b)** Its leaves are also used for external inflammation in the form of s poultice.
Diseases Cured:	Abdominal pain and external inflammation.
Ethnobotanical Uses:	Leaves are also used as fodder for cattle. Fresh leaves are cooked as a vegetable.
Phytochemicals:	Oleracein A, oleracein B, oleracein E, hesperidin, and caffeic acid [62].

3.4.55 Prunus persica (Linn.) Batsch

Family Name:	Rosaceae
Local Name/English Name:	Aru, Arro/Wild peach
Flowering Period:	March–May
Status:	Cultivated and self-growing plant
Part Used	Whole plant
Habit/Habitat:	A small tree, found as cultivated plant near houses and in waste lands.
Distribution:	***Pakistan:*** It is widely cultivated in both hilly and plain areas throughout the country. ***World:*** Cultivated in northwest Himalaya, India, Tibet, and temperate and tropical regions of the world.
Description:	A medium-sized tree up to 8 m tall. Stem erect, branched, and woody, with gray to red bark. Leaves simple, alternate, petiolate, toothed, and green, usually provided with a pair of glands at the base. Flowers pink, pedicillate, bisexual, with five sepals and five petals. Fruit drupe, fleshy and downy, enclosing a hard, furrowed, one-seeded stone (Fig. 3.73).
Medicinal Uses:	
Collection:	30–40 g of fresh leaves is collected by men and women 20–40 years old, in summer (May–September).

(continued)

Fig. 3.73 *Prunus persica* (Linn.) Batsch

Family Name:	Rosaceae
Recipes:	25–30 g of fresh leaves is ground daily for 5–6 min; 1–2 tsp (10–12 g) of common salt and 1 cup (250 mL) of water are added in. The mixture is filtered with a cloth and given to cattle to kill intestinal worms and also applied to wounds to remove maggots. Half a cup of juice (125 mL) (at one time) is given twice daily for 1–2 days. To remove maggots from wounds, 3–4 tsp (30 mL) of extract (at one time) is applied to wounds 3–4 times per day for 2–3 days.
Diseases Cured:	To kill intestinal worms, and to remove maggots from wounds in cattle and dogs.
Ethnobotanical Uses:	Leaves are also used as fodder for cattle, goats, and sheep. Fruit is edible. Unripe fruit is used in chutneys. The plant is used as firewood when dry.
Phytochemicals:	Fixed oil, called persic oil, contains B-sitosterol, and squalene, D-glucoside, hentriacontane, hentricontanol and the flavonoids naringenin, dihy-drokaempferol, kaempferol, and quercetin [1, 63].

3.4.56 *Punica granatum Linn.*

Family Name:	Punicaceae
Local Name/ English Name:	Drunni, Desi anar, Anangorai/Pomegranate
Flowering Period:	April–June
Status:	Common
Part Used:	Whole plant
Habit/Habitat:	A small tree, mostly found in waste places along slopes of lower hills in fertile loamy soil.
Distribution:	***Pakistan:*** Hazara, Salt Range, Balouchistan, Waziristan, Kurram, Dir, Chitral, and Kashmir. ***World:*** South Europe, Central and West Asia.
Description:	A small tree or shrub up to 6 m tall. Stem is erect, branched, and woody with dark gray bark. Leaves are simple, opposite, entire, oblanceolate, shining above, bright green, beneath, with spines at base. Flowers are solitary, showy, persistant, orange red with many stamens. Fruit is subglobose with leathery skin, pale red, crowned with the persistant calyx lobes and whithered stamens, and with red juicy seeds (Fig. 3.74).
Medicinal Uses:	
Collection:	4–5 kg of fresh roots, 1 kg of fresh rind, 3 kg of fresh seeds, and 1 kg of fresh flowers are collected by men, women, and children 12–40 years old. Roots are collected in winter (November–March). Seeds and rind are collected in summer (June–September). Flowers are collected in early summer (March–May). Roots are used in fresh form, while seeds, rind, and flowers are dried first in sunlight and shade for 4–6 days and then used; also stored in cotton sacks for further use.
Recipes:	**(a)** 3–4 kg of fresh roots is cleaned, washed in water 2–3 times, and cut into small 2–3-in. pieces. Next, they are boiled in 5 L of water for 1 h. When 1½ L of water is left, then it is filtered with a cloth, stored in a glass bottle, and given to patients suffering from piles and to kill intestinal worms. **For children,** 1 cup (250 mL) of decoction (at one time) is given once daily, in the morning, for 3–4 days. **For adults,** 1 cup (250 mL) of decoction (at one time) is given once per day, in the morning, for 7–8 days.

(continued)

Fig. 3.74 *Punica granatum* Linn

Family Name:	Punicaceae
	(b) 250 g of dried rind is ground for 10–15 min; 150 g of sugar is mixed in. This powder is stored in a plastic or glass bottle and given to patients suffering from diarrhea, dysentery, piles, diabetes, sore gums, stomach disorders; liver, intestinal, or bladder inflammation; and toothache. **For children,** 1 tsp (5–6 g) of powdered drug (at one time) is given with 1 cup (250 mL) of water 2–3 times per day for 4–5 days. **For adults,** 2–3 tsp (12–15 g) of powdered drug (at one time) is given with 1 cup (250 mL) of water 2–3 times per day for 8–10 days. For sore gums and toothache, 1 tsp (4–6 g) of powdered drug (at one time) is rubbed on teeth 2–3 times per day for 8–10 days. For whooping cough and sore gums, 1 tsp (4–5 g) of powdered drug or dried rind (at one time) is crushed under teeth once daily at bedtime for 3–4 days.

(continued)

Family Name:	Punicaceae
	(c) 250 g of dried flowers is ground for 8–10 min and then boiled in ½ L of water for 15–20 min; 150 g of sugar is also added. When 3 cups (750 mL) of water remains, it is filtered with a cloth, stored in a glass bottle, and given to patients suffering from stomach, liver, intestinal, or bladder inflammation, toothache, and to remove spots from face. **For children,** not used. **For adults,** 2–3 tsp (20 mL) of decoction (at one time) is given 2–3 times per day for 8–10 days. For toothache, 1 tsp (6–8 g) of powdered drug (at one time) is rubbed 2–3 times per day on aching tooth for 8–10 days. For facial spots, 1 tsp (5 mL) of decoction is rubbed on spots with cotton twice daily (morning–evening) for 10–15 days.
	(d) 1 kg of dried seeds and 250 g of sugar are ground together for 15–20 min. This powder is stored in a glass or plastic bottle and given to patients suffering from indigestion; liver, stomach, intestinal, or bladder inflammation; jaundice, fever, vomiting, diarrhea, dysentery, cold, gas trouble, and stomach disorder; also used to purify blood. **For children,** 2–3 tsp (15–20 g) of powdered drug (at one time) is mixed in 1½ cup (375 mL) of water and given 2–3 times per day for 8–10 days. **For adults,** 3–4 tsp (40–50 g) of powdered drug (at one time) is given with 1 cup (250 mL) of water 2–3 times per day for 10–15 days.
Diseases Cured:	Diarrhea, dysentery, piles, diabetes, intestinal worms, fever, whooping cough, cooling, indigestion, stomach disorder, jaundice, vomiting, sore gums, toothache, liver, stomach, bladder, or intestinal inflammation; also used to purify blood.
Ethnobotanical Uses:	Leaves are used as fodder for goats and sheep. Wood is also used to make light furniture, agricultural implements, tool handles, and handicrafts, and as fuelwood, mud roof thatching; plant used for hedges and fencing. The seeds are used in sharbat, chutneys, and curries, as spices and flavoring agent; seeds are edible.
Phytochemicals:	Tannin, punico-tannic acid, mannite sugar, gum, pectin, ash, alkaloid pelletierine, oil liquid isopelletierine, inactive alkaloids methyl-pelletierine, pseudo pelletierine, citric acid, sorbitol, mannitol, pelletierine, isoquercetrin, B-sitosterol, friedelin, D-mannitol, estrone, glucose, fructose, sucrose, maltose, oxalic acid, organic acid [1, 25].

Fig. 3.75 *Quercus incana* Roxb

3.4.57 Quercus incana Roxb.

Family Name:	Fagaceae
Local Name/ English Name:	Rain, Serai, Shah baloot/Bluejack oak
Flowering Period:	May–August
Status:	Common
Part Used:	Whole plant
Habit/Habitat:	A slowgrowing tree found mostly in colder shady and humus places.
Distribution:	***Pakistan:*** Swat, Hazara, Dir, Kashmir, Rawalpindi, Murree, and Gilliyat. ***World:*** Found in outer and temperate Northwest Himalaya to Nepal, Burma, Afghanistan, and Himalaya from Swat and Kashmir to Bhutan and China.
Description:	A tree up to 6–18 m tall. Shoots puberulous to tomentose. Leaves elliptic–lanceolate to ovate–lanceolate, serrate but not at the base, acuminate. Upper surface of the leaves is dark green, lower white tomentose. Male flowers in catkins, bract ovate–oblong to oblong–orbiculate. Nut 1–2 cm long, canescent, umbo-tipped, glabrescent (Fig. 3.75).
Medicinal Uses:	
Collection:	Up to 500 g of bark and fruit is collected by men 25–40 years old, in winter (November–March). The bark is cut into small 2–4-in. pieces and dried in shade for 10–15 days.

(continued)

Family Name:	Fagaceae
Recipes:	**(a)** Dried bark is first cut into small pieces and boiled in milk until it becomes gelatinous. This gel is used as a bandage for broken and cracked bones. **(b)** Dried bark is ground into fine powder and mixed with flour and water. This mixture is placed externally on wounds. **(c)** The bark is also kept in water and used as a dye for black-colored hair. **(d)** Similarly, the fruit's endosperm is ground, mixed with flour and water, and used to cure dysentery and diarrhea. **For children,** 1–2 tsp is given 1–2 times per day for 2–3 days. **For adults,** 1–2 cups (250–500 ml) are used for 4–5 days.
Diseases Cured:	Dysentery and diarrhea, wound healing, broken and cracked bones.
Ethnobotanical Uses:	Leaves are used as fodder for goats. Wood is used as fuel, in construction, and in making tool handles.
Phytochemicals:	Tannins, protein, cellulose, and carbohydrates [27].

3.4.58 *Quercus leucotrichophora A. Camus*

Family Name:	Fagaceae
Local Name/ English Name:	Rein, Shah baloot, Shindar/Oak
Flowering Period:	April–May
Status:	Common
Part Used:	Whole plant
Habit/Habitat:	A tall tree, mostly grows on shady slopes of hilly areas in loamy soil.
Distribution:	***Pakistan:*** Hazara, Murree, Dir, Swat, Poonch, Sakeser, and Salt Range. ***World:*** India, Nepal, and Burma.
Description:	An evergreen tree up to 20 m tall. Stem erect, branched, woody, and hard with dark gray bark. Leaves simple, elliptic–lanceolate, alternate, petiolate, and spiny-toothed with upper surface dark green and under surface white tomentose. Inflorescence catkin, long, with 4–6 stamens. Cupules broad, half-enclosing the oblong gray nut fruit (Fig. 3.76).

(continued)

Fig. 3.76 *Quercus leucotrichophora* A. Camus

Family Name:	Fagaceae
Medicinal Uses:	
Collection:	300 g of fresh bark is collected by men 25–40 years old, in winter (November–March). The bark is cut into small 2–4-in. pieces.
Recipes:	**(a)** 300 g of fresh bark of *Quercus incana* and 500 g of fresh roots of *Rumex hastatus* (Kathimal) are boiled together in 5 L of water for 2–3 h. When 1 L of water is left, it is strained with a piece of cloth. This decoction is mixed with 1/2 Kg wheat flour (*sooji*), 200 g of sugar, and 250 g of ghee (*desi*) and cooked for 20–25 min. This sweet meal (*halwa*) is given to patients suffering from asthma, cough, fever, and rheumatism. **For children,** 1–2 tsp (15–20 g) of *halwa* (at one time) is given once or twice daily for 4–5 days. **For adults,** 2–3 tsp (40–50 g) of *halwa* (at one time) is given twice daily for 10–15 days. This drug is also given to cattle suffering from body weakness; 200 g of drug (at one time) is given to cattle 2–3 times per day for 6–7 days.
	(b) 150 g of fresh bark is boiled daily in 2 cups (500 mL) of water for 25–30 min. When 1 cup (250 mL) of water is left, it is strained with a piece of cloth. This decoction is given to patients suffering from backache. **For children,** not used. **For adults,** 1 cup (250 mL) of decoction (at one time) is given once daily, before breakfast, for 6–7 days.

(continued)

Family Name:	Fagaceae
Diseases Cured:	Asthma, cough, fever, rheumatism, and backache.
Ethnobotanical Uses:	Leaves are used as fodder for cattle, goats, and sheep. Wood is used to make agricultural implements, plows, and tool handles; for thatching; and as fuelwood and timber.
Phytochemicals:	Fatty, oleic, palmatic, and linoleic acids, moisture, ash, oil, tannins, protein, cellulose, and carbohydrates [27, 64].

3.4.59 Ranunculus laetus Wall. ex Hook. F. & Thoms.

Family Name:	Ranunculaceae
Local Name/ English Name:	Chambel booti, Chaho/Buttercup
Flowering Period:	April–May
Status:	Common
Part Used:	Leaves
Habit/Habitat:	A perennial herb, commonly found in moist places along waterways.
Distribution:	***Pakistan:*** Hazara, Waziristan, Khurrm, Chitral, Swat, Gilgit, Baltistan, Murree, and Kashmir. ***World:*** Afghanistan, east to Tibet, India, and Southwest China.
Description:	A small perennial herb up to 1 ft tall. Roots branched, herbaceous, and adventitious. Stem branched, rhizome, dark brown. Leaves are simple, alternate, petiolated, green, developed on main stem. Basal leaves hairy, suborbicular, deeply two- to three-lobed; lobes are coarsely toothed. Flowers bright yellow and broad. Fruit single-seeded, achene, suborbicular (Fig. 3.77).
Medicinal Uses:	
Collection:	70–80 g of fresh leaves is collected daily by men and women 20–40 years old, in summer (April–July).
Recipes:	70–80 g of fresh leaves are ground daily for 7–8 min; 2–3 tsp (15–25 mL) of water is also added. This paste (*malum*) is applied on infected parts and wounds; 1–2 tsp (12–15 g) of paste (at one time) is applied on infection and covered with cloth; or 2–3 tsp (20–30 g) is applied to wound once per day for 1–2 days.
Diseases Cured:	Skin infections (*chambal*); to kill germs in dog and cattle wounds.
Phytochemicals:	Poisonous substance glycoside ranunculin [3].

Fig. 3.77 *Ranunculus laetus* Wall. ex Hook. F & Thoms

3.4.60 Rhus chinensi L.

Family Name:	Anacardiaceae
Local Name/ English Name:	Titar, Tatrak, Sumak/Sumac
Flowering Period:	March–July
Status:	Rare
Part Used:	Whole plant
Habit/Habitat:	A dioecious tree, grows in waste, shady, and cold places in clay loam.
Distribution:	***Pakistan:*** Swat, Hazara, Murree Hills, Margallah Hills, and Kashmir. ***World:*** Afghanistan, India, and western Nepal.
Description:	A small dioecious tree up to 4 m tall. Stem erect, branched, and woody, with grayish bark. Leaves compound, 10–12 leaflets, petiolate, opposite, lanceolate, toothed, upper surface dark green, under surface gray to whitish. Flowers unisexual, numerous, small, white to creamy, in clusters. Fruit drupe, small, rounded globose, in clusters, single-seeded (Fig. 3.78).

(continued)

Fig. 3.78 *Rhus chinensis* L

Family Name:	Anacardiaceae
Medicinal Uses:	
Collection:	2 kg of fresh fruit and 1 kg of roots are collected by men 20–35 years old, in summer (July–September). Roots are used fresh, whereas fruit is dried in shade for 8–10 days. Roots are cleaned and washed in water 2–3 times and cut into small 1–2-in. pieces.
Recipes:	**(a)** 1 kg of dried fruit is boiled in 6–7 cups (1,500–2,000 mL) of water for 25–30 min; when 3–4 cups (750–1000 mL) of water is left, it is filtered with a cloth. Next, 300 g of sugar is mixed into this filtrate and boiled again for 10–15 min. Then it is put in a glass bottle and given to patients suffering from jaundice, asthma, and cough. **For children,** 1–2 tsp (10–15 mL) of decoction is mixed in 1 cup (250 mL) of water (at one time) and given 2–3 times per day for 14–15 days. **For adults,** 3–4 tsp (25–35 mL) of decoction is mixed in 2 cups (500 mL) of water (at one time) and given 3–4 times per day for 25–30 days.

(continued)

Family Name:	Anacardiaceae
	(b) 250 g of dried fruit of *Rhus chinensis*, 250 g of dried seed of *Punica granatum*, 250 g of dry rind of *Punica granatum*, 250 g of dried *Foeniculum vulgare*, and 300 g of sugar (*gur*) are mixed together and ground for 25–30 min. This powder is then stored in a glass or plastic pot and given to patients suffering from asthma, jaundice, and body inflammation. **For children,** 1–2 tsp (8–12 g) of powdered drug (at one time) is given with 1 cup (250 mL) of water 2–3 times per day for 10–15 days. **For adults,** 3–4 tsp (20–25 g) of powdered (at one time) is given with 1 cup (250 mL) of water 2–3 times per day for 20–25 days.
	(c) 1 kg of fresh roots is cleaned, washed 2–3 times with water, and cut into small 1–2-in. pieces. Then the pieces are boiled in 6–8 cups (1500–2000 mL) of water for 20–30 min. When 3–4 cups (750–1000 mL) of water is left, it is filtered with a cloth. This decoction is stored in a glass bottle and given to patients suffering from jaundice. **For children,** 1–2 tsp (10–12 mL) of decoction (at one time) is mixed in ½ cup (125 mL) of water and given 2–3 times per day for 10–15 days. **For adults,** 3–4 tsp (25–30 mL) of decoction (at one time) is mixed in 2 cups (500 mL) of water and given 2–3 times per day for 25–30 days.
Diseases Cured:	Jaundice, cough, asthma, and body inflammation.
Ethnobotanical Uses:	Leaves are used as fodder by goats and sheep. The plant is also used as firewood when dry.
Phytochemicals:	Tannins, raffinose, sucrose, glucose, fructose, rhamnose, myricetin, quercetin, kaemferol, gallic acid, methyl gallate, ellagic acid, gallotannin, gallic acid, benzofuran lactone, rhuscholide A, 5-hydroxy-7-(3,7,11,15-tetramethylhexadeca-2,6,10,11-tetraenyl)-2(3*H*)-benzofuranone, betulin, betulonic acid, moronic acid, 3-oxo-6β-hydroxyolean-12-en-28-oic acid, and 3-oxo-6β-hydroxyolean-18-en-28-oic acid [1, 65].

Fig. 3.79 *Rosa brunonii* Lindley

3.4.61 Rosa brunonii Lindley

Family Name:	Rosaceae
Local Name/ English Name:	Tarni, Jangli gulab, Gulab gul/Wild rose
Flowering Period:	April–June
Status:	Uncommon
Part Used:	Whole plant
Habit/Habitat:	A prickly and spiny climbing shrub.
Distribution:	***Pakistan:*** Balouchistan, Kurram, Chitral, Swat, Hazara, Gilgit, Murree, Poonch and Kashmir. ***World:*** Burma, Bhutan, India and west China.
Description:	A climbing shrub up to 6 m tall. Stem climber, branched, woody, and prickly. Leaves compound, alternate, 5–9 leaflets, elliptic–lanceolate to elliptic–oblong, opposite, petiolate, toothed and green with imparipinnate stipules. Flowers white, in panicles, padiceliate. Fruit oval-shaped, subglobose, brown (Fig. 3.79).
Medicinal Uses:	
Collection:	50–70 g of fresh flowers is collected daily by men and women 20–40 years old, in summer (April–June).

(continued)

Family Name:	Rosaceae
Recipes:	50 g of fresh flowers is boiled in 1 cup (250 mL) of water for 10–15 min. When 1 cup (250 mL) of water remains, it is filtered with a cloth and given to patients suffering from constipation. **For children,** 1–2 tsp (10–12 mL) of decoction (at one time) is given twice daily (morning–evening) for 1–2 days. **For adults,** ½ cup (125 mL) of decoction (at one time) is given twice daily (morning–evening) for 3–4 days.
Disease Cured:	Constipation.
Ethnobotanical Uses:	Leaves and young twigs are used as fodder for goats and sheep. The plant is also used for fencing and hedges, and as firewood when dry.
Phytochemicals:	The essential oil consists mainly of eugenol (30%), citronellol (2.65%), geraniol (10.5%), and terpinen-4-ol (13.7%) as the major compounds [66].

3.4.62 *Rosa indica Linn.*

Family Name:	Rosaceae
Local Name/ English Name:	Gulab, Gulab bhool/Rose
Flowering Period:	March–August
Status:	Common
Part Used:	Whole plant
Habit/Habitat:	A prickly or spiny shrub, found as a cultivated plant in home gardens. It is propagated by its vegetative part (branches).
Distribution:	***Pakistan:*** Found throughout country as common home garden and park cultivated plant. ***World:*** Distributed throughout the world.
Description:	A small prickly or spiny shrub up to 1 m tall. Stem erect, branched, woody, and prickly. Leaves compound with 3–5 leaflets, toothed, opposite, petiolate, and green. Flowers solitary, pinkish, red, and white. Fruit subglobose, fleshy, and bright red (Fig. 3.80).
Medicinal Uses:	
Collection:	2 kg of fresh flowers is collected by men, women, and children 12–40 years old, in summer (April–August).

(continued)

Fig. 3.80 *Rosa indica* Linn

Family Name:	Rosaceae
Recipes:	1 kg of fresh flower petals and 1 kg of sugar are crushed together by hand for 20–25 min and then placed in sunlight for 7–8 days. A mixture, "*gulkand*," is prepared after 1 week, stored in a glass jar, and given to patients suffering from constipation, inflammation, leucorrhoea, heart and eye diseases. **For children,** 1 tsp (5–6 g) of drug (at one time) is given 2–3 times per day for 2–3 days. **For adults,** 2–3 tsp (25–30 g) of drug (at one time) is given 2–3 times per day for 15–20 days.
Diseases Cured:	Constipation, body inflammation, heart and eye diseases, and leucorrhoea.
Ethnobotanical Uses:	It is also cultivated in home gardens as an ornamental plant because of its beautiful flowers.
Phytochemicals:	Essential oil, phenolic antioxidants, hydrolyzable tannins, flavonols, and anthocyanins [30, 67].

3.4.63 Rumex dentatus Linn.

Family Name:	Polygonaceae
Local Name/ English Name:	Hula, Ambarvati, Taluni/Dock
Flowering Period:	March–May

(continued)

Fig. 3.81 *Rumex dentatus* Linn

Family Name:	Polygonaceae
Status:	Common
Part Used:	Whole plant
Habit/Habitat:	An erect annual herb, found as self-growing plant in waste places and in the cultivated fields of *Zea mays* in fertile soil.
Distribution:	***Pakistan:*** Hazara, Peshawar, Quetta, Ziarat, Wazirestan, Parachinar, Kashmir, Rawalpindi and Islamabad. ***World:*** Iran, Afghanista, India and China.
Description:	A small annual herb up to 2 ft tall. Stem erect, branched, herbaceous, green, provided with nodes and internodes. Leaves simple, alternate, redical, pale green, petiolate with crispy margin. Flowers small, greenish, numerous, in terminal branched racemes. Fruiting sepals enlarged, ovate or obovate, winged; wings nervosa, toothed (Fig. 3.81).

(continued)

Family Name:	Polygonaceae
Medicinal Uses:	
Collection:	½ kg of fresh roots is collected daily by men 20–40 years old, in summer (March–July), then cleaned, washed 2–3 times in water, and cut into small 1–2-in. pieces.
Recipes:	½ kg of fresh roots and ½ cup (60–80 g) of common salts are ground together for 10–15 min. This paste (*pinna*) is given to cattle suffering from diarrhea, dysentery, and intestinal worms; 250 g of drug (at one time) is given twice daily (morning–evening) for 2–3 days. The drug's taste is sour and its color is light pink. Half a liter (500 mL) of milk should be given 8–10 min before the drug is used.
Diseases Cured:	Diarrhea, dysentery, and intestinal worms in cattle.
Ethnobotanical Uses:	Young leaves are used as spinach (*sag*). The whole plant is used as fodder for cattle.
Phytochemicals:	Oxalic acid, potassium binoxalate, oxymethyl anthraquinone, tartaric acid, tannin, and vitamins [1].

3.4.64 *Rumex hastatus D. Don*

Family Name:	Polygonaceae
Local Name/ English Name:	Khatimal, Khatembul/Yellow sock
Flowering Period:	March–August
Status:	Common
Part Used:	Whole plant
Habit/Habitat:	A bushy perennial small shrub, mostly grows in waste lands on dry, hard, and sandy soil.
Distribution:	***Pakistan:*** Hazara, Tirah, Malakand, Peshawar, Chitral, Drosh, Swat, Gilgit, Murree, and Kashmir. ***World:*** Afghanistan, India, Bhutan, and West China.
Description:	A bushy small shrub up to 2 ft tall. Stem erect, branched, herbaceous above and woody below, pale green to light brown. Leaves simple, pale green, hastate; lobes are directed outward. Flowers small, numerous, pinkish in terminal paniculate clusters. Fruit pinkish and one-seeded nutlet (Fig. 3.82).

(continued)

Fig. 3.82 *Rumex hastatus* D. Don

Family Name:	Polygonaceae
Medicinal Uses:	
Collection:	½ kg of fresh roots is collected by men and women 20–40 years old, in winter (December–March), then cleaned and washed in water 2–3 times, and cut into small 1–2-in. pieces.
Recipes:	½ kg of fresh roots of *Rumex hastatus* and 250 g of fresh bark of *Quercus incana* are boiled together in 4 L of water for 1–2 h. When 1 L (4 cups) of water remains, it is filtered with a cloth; 250 g of sugar, 1 kg of coarsely ground flour (*sooji*), and 250 g of fat (*desigee*) are added to it, and the mixture is cooked for 10–15 min. This sweet meal (*halwa*) is given to patients suffering from cough, asthma, and fever. **For children,** 2–4 tsp (30–40 g) of drug (at one time) is given 2–3 times per day for 3–4 days. **For adults,** 8–10 tsp (80–100 g) of drug (at one time) is given 2–3 times per day for 10–15 days.
Diseases Cured:	Asthma, cough, and fever; and for general body weakness in cattle.

(continued)

Family Name:	Polygonaceae
Ethnobotanical Uses:	Young leaves are used in chutneys and as spinach (*sag*) to improve taste, and as a flavoring agent. The plant is also used as fodder by cattle, goats, and sheep.
Phytochemicals:	Nepalin, nepodin, rumicin, hastatusides A and B (1 and 2, resp.), resveratrol, rumexoside, torachrysone-8-yl β-D-glucopyranoside, rutin, and orientaloside [27, 68].

3.4.65 *Salvia moorcroftiana* Wall. ex Benth

Family Name:	Lamiaceae
Local Name/ English Name:	Kaljari, Gahi kand, Lupra/Wild sage
Flowering Period:	March–April
Status:	Common
Part Used:	Roots
Habit/Habitat:	A small herb, found in waste places alongside cultivated fields in clay loam.
Distribution:	***Pakistan:*** Quetta, Waziristan, Kurram, Hazara, Salt Range, Taxila, Dir, Chitral, Swat, Kashmir, Jhelum, Rawalpindi, and Islamabad. ***World:*** India, Afghanistan, Bhutan, Nepal, and Central Asia.
Description:	A small herb up to 1 ft tall. Stem erect, branched, herbaceous, provided with nodes and internodes, hairy. Leaves simple, petiolate, whorled, toothed, with upper surface green, and white under surface, hairy. Flowers violet, blue, and small (Fig. 3.83).
Medicinal Uses:	
Collection:	250 g fresh roots is collected by men and women 20–40 years old, in summer (June–August), cleaned, washed in water 2–3 times, and then dried in sunlight for 3–4 days.
Recipes:	200 g of dried roots is ground for 8–10 min. This powder is then stored in a glass or plastic bottle and given to patients suffering from gas trouble, stomach disorders, diarrhea, and cough. **For children,** 1 tsp (4–6 g) of powdered drug (at one time) is given with 1 cup (250 mL) of water 2–3 times per day for 3–4 days. **For adults** (16–60 years), 2–3 tsp (15–20 g) of powdered drug (at one time) is given 2–3 times per day for 4–5 days with 1 cup (250 mL) of water.

(continued)

Fig. 3.83 *Salvia moorcroftiana* Wall. ex Benth

Family Name:	Lamiaceae
Diseases Cured:	Diarrhea, gas trouble, stomach disorders, and cough.
Phytochemicals:	Volatile oil containing salven, pinene, camphor, cineole, bromeol, thujone, salvene ester, and sequiterpenes [69].

3.4.66 *Sageretia brandrethiana Aitch., J.L.S.*

Family Name:	Rhamnaceae
Local Name/ English Name:	Ganger, Gangern/Mock Buckthorn
Flowering Period:	March–August
Status:	Common
Part Used:	Whole plant
Habit/Habitat:	A spreading shrub, mostly found in waste places on lower hills in dry and hard soil.

(continued)

Fig. 3.84 *Sageretia brandrethiana* Atich., J.L.S

Family Name:	Rhamnaceae
Distribution:	***Pakistan:*** Ziarat, Hazara, Peshawar, Kurram, Dir, Swat, Chitral, Bundi, Berikot, Muzfarabad, Salt Range, Jhelum, Rawalpindi district, and Margalla Hills. ***World:*** India, Sri Lanka and Tropical and sub tropical North Africa eastward to south Iran, Greece and Afghanistan
Description:	A small shrub up to 2 m tall. Stem erect, branched, and woody, with grayish bark. Leaves simple, small, ovate, entire, rounded at the apex, tapering toward the base; upper surface is dark green and under surface is whitish. Flowers yellowish, numerous, small. Fruit rounded and black (Fig. 3.84).
Medicinal Uses:	
Collection:	2 kg of fresh roots and 1 kg of fresh leaves are collected by men, women, and children 14–40 years old. Roots are collected in winter (November–February), cleaned, washed in water 2–3 times, and cut into small 2–3-in. pieces. Leaves are collected in summer (March–July), dried in shade for 5–6 days, and stored in cotton sacks for further use.

(continued)

Family Name:	Rhamnaceae
Recipes:	**(a)** 2 kg of fresh roots is boiled in 7–8 L of water for 2–3 h. When 2 L of water is left, it is filtered with a cloth and stored in a glass bottle. The decoction is given to patients suffering from cough, asthma, jaundice, kidney stones, and body weakness. **For children,** ½ cup (125 mL) of decoction (at one time) is mixed with 1 cup (250 mL) of water and given twice daily (morning–evening) for 7–8 days. **For adults,** 1 cup (250 mL) of decoction (at one time) is mixed with 1 cup (250 mL) of water and given twice daily (morning–evening) for 8–10 days. **(b)** 120 g of dried leaves and roots is ground together for 8–10 min. This powder is stored in a glass or plastic bottle and given to patients suffering from diarrhea, dysentery, and skin diseases. **For children,** ½ tsp (3–4 g) of the drug (at one time) is given with 1 cup (250 mL) of water 2–3 times per day for 3–4 days. **For adults,** 1 tsp (6–8 g) of the drug (at one time) is given with 1 cup (250 mL) of water for 5–6 days.
Diseases Cured:	Cough, asthma, jaundice, dysentery, diarrhea, skin diseases, kidney stones, and general body weakness.
Ethnobotanical Uses:	Leaves are used as fodder for goats and sheep. The fruit is edible. The plant is used in mud roof thatching and as firewood.

3.4.67 Solanum nigrum L.

Family Name:	Solanaceae
Local Name/ English Name:	Kachmach, Mako/Common nightshade
Flowering Period:	March–December
Status:	Uncommon
Part Used:	Whole plant
Habit/Habitat:	A small herb, grows in waste places near houses and alongside cultivated fields and roadsides in loamy soil.
Distribution:	***Pakistan:*** Found everywhere in waste places. ***World:*** Cosmopolitan.

(continued)

Fig. 3.85 *Solanum nigrum* L

Family Name:	Solanaceae
Description:	A small herb up to 2 ft tall. Stem erect, branched, herbaceous above, woody below, and green. Leaves simple, opposite, petiolate; oval-shaped, hairy, green, and developed on main stem and branches. Flowers white. Fruit rounded berry, orange-red to black with many white seeds (Fig. 3.85).
Medicinal Uses:	
Collection:	125 g of fresh leaves is collected by men, women, and children 12–40 years old, in summer (April–August).
Recipes:	125 g of fresh leaves is boiled in 3–4 cups (750–1000 mL) of water daily for 10–15 min. When 1 cup (250 mL) of water remains, then it is filtered with a piece of cloth; used to treat swelling, skin diseases, inflamed and painful parts of body and to clean wounds and mouth sores. Half a cup (125 mL) of decoction (at one time) is applied 3–4 times per day for 6–7 days for mouth sores.
Diseases Cured:	Body and joint swelling, skin diseases, mouth sores, inflamed and painful body parts; to clean wounds.
Ethnobotanical Uses:	Young leaves are used as spinach (*sag*) and also used as fodder by goats, sheep, and cattle. Berries are edible.

(continued)

Fig. 3.86 *Solanum surratense* Burm, f

Family Name:	Solanaceae
Phytochemicals:	Alkaloidal glycosides, solasonine, solamargine, B-solamargine, tigogenin, solasodine a-solasonine, saccharopine, 2-aminoadipic acid, and hemagglutinins [70].

3.4.68 Solanum surattense Burm. f.

Family Name:	Solanaceae
Local Name/ English Name:	Mohri, Kateli, Maraghona/Indian solanum
Flowering Period:	March–December
Status:	Uncommon
Part Used:	Whole plant
Habit/Habitat:	A branched herb, grows mostly in waste places in dry rough clay.
Distribution:	***Pakistan:*** Found in all four provinces. ***World:*** North Africa, South and Southeast Asia, Australia, and Polynesia.
Description:	A prostrate, branched, prickly herb up to 1 m in length. Stem prostrate, herbaceous, dark green, with yellow thorns and branches. Leaves simple, dark green, petiolate, elliptic–oblong, wavy, with yellow thorns. Flowers bluish purple, 2–4 in number, with yellow anthers. Fruit yellow, globose berry with many seeds (Fig. 3.86).

(continued)

Family Name:	Solanaceae
Medicinal Uses:	
Collection:	250 g of fresh plant material is collected by men and women 20–40 years old, in summer (March–July). First, thorns are removed by keeping the plant on a light fire. Then it is cut into small 1–2-in. pieces.
Recipes:	200 g of plant material is boiled in 1 L of water daily for 20–25 min. Five to six small amount black peppers and 2–3 tsp (20–25 g) of common salt is also added. When ½ L of water remains, it is filtered with a cloth or filtration pot. This decoction is given to cattle and patients suffering from indigestion, fever, cough, stomach disorders, and asthma and to improve hunger. **For children,** ½ cup (125 mL) of decoction (at one time) is given with bread 2–3 times per day for 6–7 days. **For adults,** 1–2 cups (250–500 mL) of decoction (at one time) is given with bread 2–3 times per day for 8–10 days. **For cattle:** 1 cup (250 mL) of decoction if given (at one time) twice daily (morning–evening) for 4–5 days.
Diseases Cured:	Fever, cough, stomach disorders, asthma, and indigestion in cattle.
Ethnobotanical Uses:	The plant is used as fodder by camels. It is also cooked as a vegetable.
Phytochemicals:	Agluco-alkaloid solancarpine, solacarpidine, and sterol "Carpesterol" [28].

3.4.69 *Tagetes minuta Linn.*

Family Name:	Asteraceae
Local Name/ English Name:	Sadbarga/Marigold
Flowering Period:	May–December
Status:	Common
Part Used:	Leaves
Habit/Habitat:	A wild herb, grows on waste places alongside roads and cultivated fields in clay.
Distribution:	***Pakistan:*** Hazara, Murree Hill, Margalla Hills, Rawalpindi district, Salt Range, and Kashmir. ***World:*** Found in tropical and temperate regions of the world.

(continued)

Fig. 3.87 *Tagetes minuta* Linn

Family Name:	Asteraceae
Description:	A small annual herb up to 1 m tall. Stem erect, branched, and herbaceous, with red skin. Leaves compound; arise from main stem or branches; leaflets, opposite, sessile, with toothed margin, 8–12 in number. Flowers small, yellow, numerous, in clusters. Fruit small, in groups (Fig. 3.87).
Medicinal Uses:	
Collection:	150 g of fresh leaves is collected by men, women, and children 12–40 years old, in winter (August–December).
Recipes:	150 g of fresh leaves are ground daily for 2–3 min. This paste (*malum*) (40–50 g) (at one time) is applied to wounds 2–3 times per day for 3–4 days.
Diseases Cured:	To kill germs and for wound healing.
Phytochemicals:	Volatile oil consists of d-limonene, ocimene, ocimenone, tagetone, dihydrotagetone, and 2, 6-dimethyl-7-octen-4-one [36].

Fig. 3.88 *Taraxacum officinale* Webber

3.4.70 *Taraxacum officinale Weber*

Family Name:	Asteraceae
Local Name/ English Name:	Hand, Kanphool, Zear gulay, Dudal/Dandelian
Flowering Period:	February–May
Status:	Common
Part Used:	Whole plant
Habit/Habitat:	A common perennial herb, mostly grows in open waste places alongside cultivated fields in clay loam.
Distribution:	***Pakistan:*** Found in all four provinces. ***World:*** Cosmopolitan.
Description:	A small, glabrous, perennial herb with rosette of basal leaves. Stem underground, long, simple or branched rhizome. Plant parts exude a milky juice. Leaves arise in rosette from rhizome, bright green, pinnately lobed, and runcinate-lyrate. Flowers yellow, all ligulate, born on scapigerous heads. Fruit achene, greenish-brown, terminated by cylindrical stalk, which is surrounded by a ring of hairs (pappus) (Fig. 3.88).

(continued)

Family Name:	Asteraceae
Medicinal Uses:	
Collection:	½ kg of fresh rhizomes is collected by men and women 25–40 years old, in summer (June–October). These rhizomes are cleaned, washed with water (2–3 times), cut into small 1–2-in. pieces, and then dried in sunlight for 2–3 days. They are stored in cloth or a mud pot for further use.
Recipes:	10–15 g of dried rhizomes are boiled daily in 3 cups (750 mL) of water for 15–16 min and 15–20 g of sugar is added. When 2 cups (500 mL) of water is left, it is strained with a cloth. This decoction is given to patients suffering from jaundice. **For children,** ½ cup (125 mL) of decoction (at one time) is given twice daily(morning–evening) for 6–7 days. **For adults,** 1 cup (250 mL) of decoction (at one time) is given twice daily (morning–evening) for 10–15 days.
Disease Cured:	Jaundice.
Ethnobotanical Uses:	Young leaves are cooked as spinach (*sag*). The plant is also grazed by cattle, goats, and sheep.
Photochemicals:	Taraxacin, inulin, resin, taraxacerin, levulin, mucilage, pectin, sugar, cholin, taraxatirol, mannite, essential oil, gum, Ca-salt, albumen, glutin, provitamins A, B, C and D; tetrahydroridentin B, taraxacolide-*O*-β-glucopyranoside, guaianolides 11β,13-dihydrolactucin, ixerin D, esters, taraxinic acid β-glucopyranoside, 11,13-dihydro-derivative, ainslioside, taraxacoside, an acylated γ-butyrolactone glycoside, various triterpenes and phytosterols such as taraxasterol, ψ-taraxasterol, their acetates and their 16-hydroxy derivatives arnidol and faradiol, α- and β-amyrin, β-sitosterol, β-sitosterol-β-d-glucopyranoside, and stigmasterol [28, 71].

3.4.71 Tribulus terrestris L.

Family Name:	Zygophylaceae
Local Name/ English Name:	Markondai/Small caltrop
Flowering Period:	April–August
Status:	Common

(continued)

Fig. 3.89 *Tribulus terrestris* L

Family Name:	Zygophylaceae
Part Used:	Fruit
Habit/Habitat:	A small herb, found in sandy soil at barren places.
Distribution:	***Pakistan:*** Gilgit, Chitral, Dir, Swat, Hazara, Kashmir, Baltistan, Jhelum, Sargodha, Rawalpindi, Multan, Lahore, D. G. Khan, Khairpur, Karachi, Thatta, Tharparkar, Dadu, and Quetta. ***World:*** Tropical and subtropical countries in Asia, Africa, S. Europe, N. Australia, and introduced to tropical regions of the New World.
Description:	Annual or biennial, prostrate, whitish silky pubescent herb. Stem hirsute to sericeous, branches spreading. Leaves paripinnate, leaflets ovate to elliptic–oblong, broad. Flowers yellow. Fruit mericarps, densely crested and tuberculate on dorsal side, densely hairy to glabrescent, spiny (Fig. 3.89).
Medicinal Uses:	
Collection:	½ kg of fresh fruit is collected by men or women 20–40 years old, daily in September and October. Fruit are cleaned, washed, and dried under shade up to 1 week.

(continued)

Family Name:	Zygophylaceae
Recipes:	**(a)** 200 g of dried fruit of *Tribulus* is ground into powder and boiled in water. This decoction is used to cure dorsal pain, for body cooling, and as a tonic. **For children,** 1 cup (250 mL) of drug twice daily is given for 10–15 days. **For adults,** 1 cup (250 mL) of drug is given 2–3 times per day for 20–25 days.
Phytochemicals:	Terrestribisamide, 25*R*-spirost-4-en-3, 12-dione and tribulusterine, *N*-*p*-coumaroyltyramine, terrestriamide, hecogenin, aurantiamide acetate, xanthosine, fatty acid ester, ferulic acid, vanillin, *p*-hydroxybenzoic acid, and β-sitosterol [72].

3.4.72 Trichodesma indicum (L.) R. Br.

Family Name:	Boraginaceae
Local Name/ English Name:	Handusi booti, Nila karaji/Indian borage
Flowering Period:	March–August
Status:	Common
Part Used:	Whole plant
Habit/Habitat:	An annual prostrate herb, grows in waste places in grassy fields in clay.
Distribution:	***Pakistan:*** Sind, Karachi, Balouchistan, Chitral, Hazara, Swat, Jhelum, and Kashmir. ***World:*** Afghanistan, India, Philippines, and Mauritius.
Description:	A small annual, hairy, and prostrate herb up to 1 ft tall. Stem erect, branched, herbaceous, hairy, light green. Leaves simple, alternate, hairy, lannceolate to oblanceolate, entire, and pedicellate. Flowers are pinkish-blue, many, solitary, or in terminal short racemes. Anthers hairy. Fruit smooth and nutlets (Fig. 3.90).
Medicinal Uses:	
Collection:	½ kg of fresh plant is collected by men or women 20–40 years old, daily in any season; then it is cleaned and washed once or twice in water.

(continued)

Fig. 3.90 *Trichodesma indicum* (L.) R. Br

Family Name:	Boraginaceae
Recipes:	**(a)** 70–80 g fresh leaves is ground daily with 50 g of sugar for 4–5 min. Next, 1 cup (250 mL) of water is added and then filtered with a piece of cloth. This filtrate is then given to patients suffering from diarrhea, dysentery, and stomach inflammation. **For children,** 1 cup (250 mL) of drug is given once in the morning for 4–5 days. **For adults,** 1 cup (250 mL) of drug is given once in the morning for 7–8 days. **(b)** 200 g of fresh leaves and roots is ground daily for 6–7 min. This paste (*malum*) is applied to painful joints and wounds 2–3 times per day for 3–4 days. **(c)** ½ kg of fresh plant is ground daily for 10–12 min; 1 cup (250 mL) of water is mixed in. This paste (*pinna*) is given to cattle suffering from stomach disorders and to kill intestinal worm; 250 g of drug is given twice daily (morning–evening) for 2–4 days.
Diseases Cured:	Diarrhea, dysentery, stomach inflammation, joint swelling, stomach disorders; to soften wounds; and to kill intestinal worms in cattle.
Ethnobotanical Uses:	The plant is used as fodder for cattle.
Phytochemicals	Steroids, triterpenoids, and lipids [73].

Fig. 3.91 *Tylophora hirsuta* Linn

3.4.73 Tylophora hirsuta Linn.

Family Name:	Asclepidaceae
Local Name/ English Name:	Panja booti/Glow
Flowering Period:	March–May
Status:	Uncommon
Part Used:	Roots and leaves
Habit/Habitat:	Climbing herb, grows in hard, dry, sunny sites.
Distribution:	***Pakistan:*** Haripur, Abbottabad, Murree, Hazara, Chitral, Swat, and Kashmir. ***World:*** Found in India and Congo.
Description:	A climbing herb up to 2 m in length. Stem climber, branched, herbaceous above, woody below, green, and hairy. Leaves simple, opposite, petiolate, entire, hairy, dark green above and light green under surface. Flowers yellow. Fruit an etaerio of two, with many brown seeds crowned with long hairs, which help in dispersal (Fig. 3.91).
Medicinal Uses:	
Collection:	20–25 g of fresh roots is collected daily by men 20–40 years old, in late summer (August–November), cleaned, washed in water 2–3 times, and cut into small 3–4-cm pieces.

(continued)

Family Name:	Asclepidaceae
Recipes:	20 g of fresh roots is ground daily for 6–7 min. Then 1 cup (250 mL) of water is added; it is filtered with a cloth and given to patients to inhibit diarrhea and vomiting. **For children**, not used. **For adults,** 1 cup (250 mL) of drug (at one time) is given only once per day for 1–2 days.
Diseases Cured:	Diarrhea and vomiting.
Ethnobnotanical Uses:	Leaves and young stems are used as fodder by goats and sheep.
Phytochemicals:	An alkaloid tylophorine [27].

3.4.74 Verbascum thapsus Linn.

Family Name:	Scrophlariaceae
Local Name/ English Name:	Gidhar tambco, Khargnwag/Mullein
Flowering Period:	June–August
Status:	Rare
Part Used:	Leaves
Habit/Habitat:	A stout herb, found on dry rocky slopes in hard and dry clay.
Distribution:	***Pakistan:*** Balouchistan, Kurram, Dir, Chitral, Swat, Hazara, Gigit, Baltistan, Dras, and Kashmir. ***World:*** Temperate Eurasia, Afghanistan, India, and Southwest China.
Description:	A densely woody, stout perennial herb, up to 1 m tall. Stem is simple, erect, unbranched, and covered with dense tomentum. Leaves simple, sessile, winged along stem, green, covered with dense tomentum. Flowers yellowish in stout spikes with dense hairy filaments. Fruit cup-shaped, grouped together on short stalks (Fig. 3.92).
Medicinal Uses:	
Collection:	½ kg of fresh leaves is collected by men 20–40 years old, in summer (April–August).
Recipes:	½ kg of fresh leaves is ground daily for 8–10 min and then mixed with ½ kg of *Triticum aestivum* flour (*pinna*), 2–3 tsp (15–20 g) of salt, and 1 cup (250 mL) of water. This paste is given to cattle suffering from dysentery and diarrhea; 150–200 g of drug (at one time) is given twice daily (morning–evening) for 2–3 days.

(continued)

Fig. 3.92 *Verbescum thapsus* Linn

Family Name:	Scrophlariaceae
Diseases Cured:	Dysentery and diarrhea in cattle.
Phytochemicals:	Mucilage, traces of volatile oil, tannin, and wax [56, 57].

3.4.75 Viola canescens Wall. ex Roxb.

Family Name:	Violaceae
Local Name/ English Name:	Phul-naqsha, Banafsha, Bamasha/Sweet violet
Flowering Period:	March–April
Status:	Common
Part Used:	Whole plant
Habit/Habitat:	A small herb, grows in waste, cold, shady places in fertile loamy soil.

(continued)

Fig. 3.93 *Viola canescens* Wall. ex Roxb

Family Name:	Violaceae
Distribution:	***Pakistan:*** Hazara, Kurram, Dir, Chitral, Swat, Murree Hills, Galis, Kohat, Salt Range, Jhelum Valley, and Kashmir. ***World:*** Iran, Afghanistan, India, and Bhutan.
Description:	A small stoloniferous herb up to 10 cm tall. Stem underground and dark brown. Leaves pubescent, simple, petiolate, toothed, chordate, and green. Flowers pale violet or violet. Spur long, straight to slightly curved. Fruit a small capsule, hairy (Fig. 3.93).
Medicinal Uses:	
Collection:	1 kg of fresh plant material is collected by men and children 14–40 years old, in summer (March–June), and dried in shade for 4–5 days.
Recipes:	10–15 g of dried or fresh plant material is boiled in 1 cup (250 mL) of water for 8–10 min; 2 tsp (10–15 g) of sugar is also added. Then it is filtered with a cloth or filtration pot and given to patients suffering from fever, cold, cough, asthma, jaundice, headache, sore throat, or toothache. **For children,** 1 cup (250 mL) of decoction (at one time) is given once daily, at bedtime, for 3–4 days. **For adults,** 2 cups (500 mL) of decoction (at one time) is given once daily, at bedtime, for 8–10 days.
Diseases Cured:	Fever, cold, cough, asthma, jaundice, headache, toothache, and sore throat.

(continued)

Family Name:	Violaceae
Ethnobotanical Uses:	Leaves are grazed by cattle.
Phytochemicals:	Triacetonamine, saponin, odoratine, glycoside, methyl salicylic ester, violine viola-quercitrin, salicylate, gum, mucilage, and sugar [1, 30].

3.4.76 Vitex negundo Linn.

Family Name:	Verbinaceae
Local Name/ English Name:	Marwani, Maravandai, Ninna/Indian privet/Five-leaved chaste tree
Status:	Common
Part Used:	Whole plant
Habit/Habitat:	A small shrub, grows mostly in waste places alongside roads, streams, and cultivated lands.
Distribution:	***Pakistan:*** Kurram, Khyber, Swat, Hazara, Rawalpindi district, Murree Hills, Jhelum Valley, and Kashmir. ***World:*** North Africa, India, and widely naturalized.
Description:	A medium-sized shrub up to 2 m tall. Stem erect, branched, and woody with grayish hairy bark. Leaves palmately compound with 3–5 leaflets, petiolate, entire with upper green and under grayish surface due to soft gray hairs. Flowers small, fragrant, pale violet to pink-mauve in axillary panicles. Corolla is with 5 subequal lobes. Fruit drupe and subglobose with single black seed (Fig. 3.94).
Medicinal Uses:	
Collection:	125 g of dried seeds are collected by men 20–40 years old, in winter (November–February), and stored in a plastic or glass bottle for further use.
Recipes:	60 g of dried seeds of *Vitex negundo*, 30 g of *Trachyspermum ammi,* and 2–3 tsp of common salt (15–20 g) are ground together for 10–15 min. This powder is stored in a glass or plastic bottle and given to patients suffering from stomach disorders, gas trouble, and cholera. **For children,** 1 tsp (4–6 g) of drug (at one time) is given with 1 cup (250 mL) of water 2–3 times per day for 3–4 days. **For adults,** 2–3 tsp (10–15 g) of powdered drug (at one time) is given with 1 cup (250 mL) of water 2–3 times per day for 3–4 days.

(continued)

Fig. 3.94 *Vitex negundo* Linn

Family Name:	Verbinaceae
Diseases Cured:	Gas trouble, cholera, and stomach grippe.
Ethnobotanical Uses:	Twigs are used as a toothbrush (*miswak*); elastic branches are used to make rope (*sub*) and baskets. The plant is used as fuelwood when dry.
Phytochemicals:	Gluconontitol, P-hydroxy benzoic acid, 5-hydroxy isophthalic acid, 3–4 dihydroxy benzoic acid, glucoside, flavone, 5,3 dihydroxy-7, 8,4 and 5,3 dihydroxy-6, 7,4 trimethoxy flavonones, 5-hydroxy-3, 6,7,3,4 pentamethoxy flavone and 3,5 dihydroxy 3,4,6-7-tetramethoxy flavole, glucose, 5-oxy iso phthalic acid, B-sitosterol, vanillic, luteolin, amino acid, glycine, alamine, valine, and lucine [74].

Fig. 3.95 *Vitis vinifera* Linn

3.4.77 *Vitis vinifera Linn.*

Family Name:	Vitaceae
Local Name/ English Name:	Angoor, Kwar/Vine
Flowering Period:	March–May
Status:	Cultivated
Part Used:	Fruit, leaves
Habit/Habitat:	A climbing shrub, propagated through branches in home gardens.
Distribution:	***Pakistan:*** Cultivated in Balouchistan, Swat, Chitral, Biltistan, Astor, and Kashmir. ***World:*** Cultivated in tropical and temperate regions of the world.
Description:	A climbing shrub up to 25 m in length. Stem climber, weak, branched, woody with dark brown bark. Leaves simple, alternate, petiolate, palmate, upper surface green under surface grayish-green and toothed. Flowers are yellowish-green, small, numerous, and in clusters. Fruit globose, yellow with two seeds (Fig. 3.95).
Medicinal Uses:	
Collection:	120 g of fresh leaves is collected by men 20–40 years old, in summer (April–July).

(continued)

Family Name:	Vitaceae
Recipes:	120 g of fresh leaves are ground daily for 10–15 min; 2 tsp (12–15 g) of common salt and 3 cups (750 mL) of water are also added. Then it is filtered and given to patients suffering from kidney pain/stone. **For children,** not used. **For adults,** 1 cup (250 mL) of juice (at one time) is given 2–3 times per day for 4–5 days.
Disease Cured:	Kidney pain (stone).
Ethnobotanical Uses:	Leaves are used as fodder by goats and sheep. Unripe fruit is used in chutneys; ripe fruit is edible.
Phytochemicals:	Moisture, ash, acidity, reduced sugar, grape sugar, gum tannin, tartaric acid; citric and malic acids; chlorides of K and Fe; sulfate of potash; tartrate lime, magnesia, alum, iron, albumin; resins contain Ca, Mg, K, P, and Fe beside gum; sugar, fixed oil, fat, tannic acid; tannins [9] contains 7–24% alcohol [27].

3.4.78 *Withania somnifera (L.) Dunal*

Family Name:	Solanaceae
Local Name/ English Name:	Kutilal/Winter cherry
Flowering Period:	Throughout year
Status:	Common
Part Used:	Leaves
Habit/Habitat:	A medium-sized shrub, grows in dry soil.
Distribution:	***Pakistan:*** Gilgit, Swat, Khyber Agency, Peshawar, Hazara, Nowshera, Kohat, Waziristan, Tank, D. I. Khan, Attock, Jhelum, Multan, Faisalabad, Baluchistan, and Karachi. ***World:*** Canary Islands, Mediterranean region, Iran, Iraq, Syria, Turkey, Palestine, Africa, and India.
Description:	A suffruticose shrub. Branches ascending. Shoots are stellate–tomentose. Leaves elliptic–ovate to broadly ovate, acute, cuneate or oblique, entire to repand. Flowers sessile to subsessile, greenish-yellow, in axillary. Fruit berry globose, red (Fig. 3.96).

(continued)

Fig. 3.96 *Withania somnifera* (L.) Dunal

Family Name:	Solanaceae
Medicinal Uses:	
Collection:	½ kg of leaves are collected by men, women, and children 12–40 years old.
Recipes:	200 g of leaves are crushed, a little water is added, and the juice is given to cure cholera, dysentery, and as an antiemetic. **For children,** 1–2 tsp (5–10 mL) of juice is given (only once per day) for 2–3 days. **For adults,** 2–3 tsp (12–15 mL) of drug (at one time) is given 2–3 times per day for 2–3 days.
Diseases Cured:	Cholera, dysentery; used as antiemetic.
Phytochemicals:	Steroidal lactones, alkaloids, flavonoids, tannin, etc. have been identified; withanolides, sitoindosides [75].

Fig. 3.97 *Woodfordia fruticosa* (L.) S. Kurz

3.4.79 *Woodfordia fruticosa (L.) S. Kurz*

Family Name:	Lythraceae
Local Name/ English Name:	Thawi, Gul dahwa/Fire flame bush
Flowering Period:	April–May
Status:	Common
Part Used:	Whole plant
Habit/Habitat:	A spreading shrub, found on cliffs in dry clay.
Distribution:	***Pakistan:*** Hazara, Swat, Balouchistan, Sind, Saidpur Hills, Rawalpindi, Murree, Islamabad, Poonch, Mirpur, and Salt Range. ***World:*** Tropical Africa, Sri Lanka, and China.
Description:	A shrub up to 3 m tall with spreading branches. Stem erect, branched, and woody, with reddish-gray bark. Leaves simple, entire and ovate, opposite, sessile, and pale green. Flowers bractet, numerous, bisexual, orange red to bright red, in clusters of 3–7 on branches, funnel-shaped, pedicellate. Fruit capsule, more or less elliptic, with many seeds (Fig. 3.97).

(continued)

Family Name:	Lythraceae
Medicinal Uses:	
Collection:	1 kg of fresh flowers is collected by men, women, and children 12–40 years old, in summer (April–May). They are dried in shade for 1–2 days and then stored in a glass or plastic bottle for further use.
Recipes:	½ kg of dried flowers and 1 cup (250 g) sugar are ground together for 10–15 min. This powder is stored in a glass bottle and given to patients suffering from piles, diarrhea, and dysentery, and to heal wounds. **For children,** 1 tsp (5–6 g) of drug is given (at one time) with 1 cup (250 mL) of milk or water 2–3 times per day for 2–3 days. **For adults,** 2–3 tsp (12–15 g) of drug (at one time) is given with 1 cup (250 mL) of milk or water 2–3 times per day for 10–15 days. A teaspoonful (4–5 g) of powdered flowers (at one time) is sprinkled on wounds 2–3 times per day for 4–5 days.
Diseases Cured:	Piles, diarrhea, and dysentery; wound healing.
Ethnobotanical Uses:	Fresh flowers are edible and sweet in taste. Leaves and twigs yield a dye used in painting. Leaves are used as fodder by goats and sheep. Wood is used as firewood when dry.
Phytochemicals:	Phenolics, particularly hydrolysable tannins, flavonoids, octacosanol and β-sitosterol, steroid sapogenin hecogenin, *meso*-inositol, triterpenoids lupeol, betulin, betulinic acid, oleanolic acid, ursolic acid, gallic acid, ellagic acid, bergenin, norbergenin, chrysophanol-8-*O*-β-d-glucopyranoside, naphthaquinone pigment lawsone, glycosides 3-rhamnoside, 3-β-l-arabinoside, 3-*O*-α-l-arabinopyranoside, 3-*O*-β-d-xylopyranoside, 3-*O*-(6″-galloyl)-β-d-glucopyranoside, and 3-*O*-(6″-galloyl)-β-d-galactopyranoside, 3-*O*-β-d-galactoside 3-*O*-(6″-galloyl)-β-d-galactopyranoside, naringenin 7-glucoside, kaempferol 3-*O*-glucoside, pelargonidin 3,5-diglucoside, anthocyanidin pigment, cyanidin, 3,5-diglucoside; tannins like 2,3,6-tetra-*O*-galloyl-β-d-glucose, 1,2,4,6-tetra-*O*-galloyl-β-d-glucose, 1,2,3,4,6-penta-*O*-galloyl-β-d-glucose, tellimagrandin, gemin D, heterophylliin A and oenothein B, oenothein A, isoschimawalin A, new hydrolysable tannins, isoschimawalin A, woodfordins A –I, and oenothein A [76].

Fig. 3.98 *Zanthoxylum armatum* DC. Prodor

3.4.80 Zanthoxylum armatum DC. Prodr.

Family Name:	Rutaceaea
Local Name/ English Name:	Timar, Kababa, Dambara/Prickly ash
Flowering Period:	March–May
Status:	Common
Part Used:	Whole plant
Habit/Habitat:	A medium-sized shrub, found in waste places in loamy clay.
Distribution:	***Pakistan:*** Hazara, Dir, Swat, Murree Hills, Margalla Hills, and Jhelum Valley. ***World:*** India, Southwest China, Taiwan, and Philippines.
Description:	A medium-sized prickly shrub up to 5 m tall. Stem erect, branched, woody and thorny with dark green dotted bark. Leaves compound, imperipinnate, 3–4 leaflets, opposite, petiolate, aromatic, lanceolate, entire, dotted with winged ptiole and midrib. Flowers yellow, numerous, in clusters, aromatic. Stamens 6–8 and ovary is three-lobed. Fruit capsule, rounded, reddish-green, splitting into two when ripe, with single black seed (Fig. 3.98).

(continued)

Family Name:	Rutaceaea
Medicinal Uses:	
Collection:	1 kg of fresh fruit is collected by men and women 20–40 years old, in summer (June–August), then dried in sunlight for 4–5 days, and stored in cotton sacks or a plastic bottle for further use.
Recipes:	50 g of dried fruit of *Zanthoxylum armatum*, 50 g of dried leaves of *Mentha longifolia*, 30 g of dried *Trachyspermum ammi,* 2 tsp (10–15 g) of black salts are ground together for 10–15 min. This powder is stored in a glass or plastic bottle and given to patients suffering from cholera, stomach disorders, gas trouble, and indigestion. **For children,** 1 tsp (6–8 g) of powdered drug (at one time) is given with 1 cup (250 mL) of water 2–3 times per day for 3–4 days. **For adults,** 2 tsp (12–15 g) of powdered drug (at one time) is given with 1 cup (250 mL) of water 2–3 times per day for 4–5 days. **For piles,** 4–5 ft of young stem is cut with a stone (not with an iron) and used as a stick to heal piles. **For mouth sores and toothache,** 6–8 in. young twigs are cut with a knife daily and used as a toothbrush (*miswak*) for mouth sores and toothache.
Diseases Cured:	Gas trouble, cholera, stomach disorders, piles, mouth gum, toothache, and indigestion.
Ethnobotanical Uses:	Leaves are used as fodder for goats and sheep. Fruit and flowers are used in chutneys, curries, as an aromatic and a flavoring agent to improve taste. The plant is used for hedges and fencing and as firewood when dry.
Phytochmeicals:	Essential oil, bitter crystalline principle, alkaloids, lignans flavonoids, aminoacids, monoterpene, organic acid, and volatile oil [30, 77].

3.4.81 Zizyphus numalaria Linn.

Family Name:	Rhamnaceae
Local Name/ English Name:	Ber, Bera/Jujube berries
Flowering Period:	May–August
Status:	Common
Part Used:	Whole plant

(continued)

Fig. 3.99 *Zizyphus numalaria* Linn

Family Name:	Rhamnaceae
Habit/Habitat:	A thorny shrub, found mostly in waste places along cultivated fields in dry soil.
Distribution:	***Pakistan:*** Balouchistan, Kurram, Kohat, Hazara, Dir, Chitral, Swat, Kashmir, Rawalpindi district, and Margalla Hills. ***World:*** India, Nepal, and sub-Himalayan tracts.
Description:	A shrub up to 4 m tall. Stem erect, branched, and woody, with spines and dark gray bark. Leaves simple, petiolate, entire, oval-shaped, green shiny above and yellowish under surface. Flowers small, yellowish, numerous. Fruit yellowish-red, one-seeded drupe (Fig. 3.99).
Medicinal Uses:	
Collection:	1 kg of ripe fruit and 250 g of fresh leaves are collected by men and women 20–40 years old. Fruit is collected in late summer (September–November) and dried in sunlight for 3–4 days, while leaves are collected in summer (April–September).
Recipes:	**(a)** 200 g of dried fruit is ground for 8–10 min. This powder is stored in a glass or plastic bottle and given to patients suffering from constipation. **For children,** ½ tsp (3–4 g) of powdered drug (at one time) is given with 1 cup (250 mL) of water 1–2 times per day for 3–4 days. **For adults,** 2–3 tsp (15–20 g) of powdered drug (at one time) is given with 1 cup (250 mL) of water 2–3 times per day for 8–10 days.

(continued)

Family Name:	Rhamnaceae
	(b) 200 g of fresh leaves is ground daily for 8–10 min; 1 cup (250 mL) of water is mixed in. Then it is filtered with a cloth and given to patients suffering from diarrhea and skin diseases (allergy). **For children,** ½ cup (125 mL) of drug (at one time) is given twice daily (morning–evening) for 4–5 days. **For adults,** 1 cup (250 mL) of drug (at one time) is given 2–3 times per day for 4–5 days.
Diseases Cured:	Constipation, skin diseases (allergy), and diarrhea.
Ethnobotanical Uses:	Fruit in both fresh and dried forms is edible. Leaves are used as fodder for goats and sheep. The plant is used as firewood and for hedges and fencing. Fresh leaves are spread under dead bodies, so there is no smell from corpses in the summer. Wood is used to make agricultural implements.
Phytochemicals:	Alkaloid frangufdine, triterpenoids, triterpene esters, aporphine alkaloid, zizyphus, saponins I–III and jujuba saponins I–IV, ziziphin, querecetion-3-O-glucoside, diglucoside, rutinoside, rhamnetin and eriodictyol (Lvs), sugar, amino acids, minerals, volatile constituents, phospholipid, triterpene oligoglycosides A and C, acetyljujuboside B, spinosin, vicenin, apigeni (flavonoid), B-glucopyransoyl, and B-glucopyranoside [78].

3.4.82 *Zizyphus oxyphylla Edgew.*

Family Name:	Rhamnaceae
Local Name/ English Name:	Pithni/Elanai
Flowering Period:	June–September
Status:	Common
Part Used:	Whole plant
Habit/Habitat:	A large shrub to medium–sized tree, found mostly in waste places in dry clay.
Distribution:	***Pakistan:*** Hazara, Dir, Swat, Murree Hills, Margalla Hills, Rawalpindi, and Jhelum Valley. ***World:*** India, Southwest China, Taiwan, and Philippines.
Description:	A large shrub or branched tree, and woody, with reddish-brown bark and spines. Leaves simple, alternate, entire, lanceolate, and green. Flowers yellowish. Fruit bright red, oval-shaped, drupe, single-seeded (Fig. 3.100).

(continued)

Fig. 3.100 *Zizyphus oxyphylla* Edgew

Family Name:	Rhamnaceae
Medicinal Uses:	
Collection:	3–4 kg of fresh roots, 25 g of fresh bark, and 30 g of fresh leaves are collected by men and women 20–40 years old. Roots and bark are collected in winter (November–February), whereas leaves are collected mostly in summer (April–November).
Recipes:	**(a)** 3 kg of fresh roots is first cleaned, washed in water 2–3 times, and cut into small 2–3-in. pieces. They are boiled in 5–6 L of water for 1–2 h. When 1 liter of water is left, it is filtered with a cloth, stored in a glass bottle, and given to patients suffering from mouth sores, skin diseases, pimples, skin lesions, earache, eye diseases, and high blood pressure. **For children,** 1–2 tsp (10–12 mL) of decoction (at one time) is mixed in 1 cup (250 mL) of water and given 2–3 times per day for 4–5 days. **For adults,** 3–4 tsp (30–40 mL) of decoction (at one time) is mixed in 1 cup (250 mL) of water and given 2–3 times per day for 8–10 days. **(b)** 25 g of fresh bark and 30 g of fresh leaves are ground together for 5–6 min. This paste (*malum*) is applied externally on pimples, skin lesions, and tumors; 3–4 g of paste (at one time) is applied 2–3 times per day for 2–4 days.

(continued)

Family Name:	Rhamnaceae
	(c) 250 g of fresh roots of *Zizyphus oxyphylla* and 250 g of fresh roots of *Punica granatum* are boiled in 1 L of water for 40–50 min. When 3 cups (750 mL) of water remains, it is filtered with a piece of cloth and stored in a glass bottle. It is given to patients suffering from intestinal worms. **For children,** 1–2 tsp (12–25 mL) of decoction (at one time) is given 2–3 times per day for 2–3 days. **For adults**, 3–4 tsp (12–25 mL) of decoction (at one time) is given 2–3 times per day for 4–5 days.
Diseases Cured:	Intestinal worms, skin diseases, pimples, mouth sores, earache, eye diseases, and high blood pressure.
Ethnobotanical Uses:	Leaves are used as fodder for goats, sheep, and camels. The plant is used as firewood and for hedges and fencing.
Phytochemicals:	14-membered frangulanine-type cyclopeptide alkaloid and oxyphyl line-A [39].

References

1. Ikram M, Hussain SF (1978) Compendium of medicinal plants. P.C.S.I.R Laboratories, Peshawar, pp 23–147
2. Mangrio SM, Dahot MU, Leghari SM (1995) Chemical constituents of *Equisetum debile* Roxb. Pakistan J. Plants Sci 1(1):41–48
3. Cooper MR, Johnson AW (1988) Poisonous plants and fungus. Her Majesty's Stationery Office, London, p 41
4. John JW, Coppen M, Robinson J, Kaushal AN (1988) Composition of xylem resin from *Pinus wallichiana* and *P. roxburghii*. Phytochemistry 27(9):2873–2875
5. Haq I, Hussain M (1993) Medicinal plants of Mansehra. Hamdard Med 36(3):63–100
6. Kirtikar KR, Basu BD (1993) Indian medicinal plants, vol I & II. Lalit Mohan Basu, Allahabad, pp 589–590
7. Ali MS, Ahmad VU, Azhar I, Ghani KU (1998) Some medicinally important plants and their uses. Hamdard Med 41(2):96–102
8. Chen L, Chen R, Wei K (1992) Constituents of tannins from *Euphorbia prostrata* ait. Zhongguo Zhong Yao Za Zhi 17(4):225–226
9. Rehman AU, Said HM, Ahmad VU (1986) Pakistan encyclopedia Planta Medica. Hamdard Foundation Pakistan and HEJ Research Institute, Karachi, pp 15–153
10. Haq I, Rehman M (1990) Medicinal plants of upper Swat (NWFP) Pakistan. Hamdard Med 33(3):51–68
11. Ahmad S, Kalhoro AM, Kapadia Z, Badar Y (1993) "Aloe", a biologically active and potential medicinal plant. Hamdard Med 36(1):109–115
12. Miller DF (1958) Composition of cereal grains and forages. National Academy of Sciences, National Research Council, Washington, DC, Publ. 585
13. Watt JM, Brandwijk MGB (1962) The medicinal and poisonous plants of Southern and Eastern Africa, 2nd edn. E. & S. Livingstone, Ltd., Edinburgh

14. Wealth of India (1950) A dictionary of Indian raw materials, vol 1. Council of Scientific and Industrial Research, New Delhi, pp 150–165
15. Bahourn T, Neergheen VS, Aruoma OI (2005) Phytochemical constituents of *Cassia fistula*. Afr J Biotechnol 4(13):1530–1540
16. Hussein FTK (1983) Medicinal plants in Libya. Arab Encylopedia House, Beirut, pp 7–822
17. Khan H, Saeed M, Gilani AUH, Khan MA, Dar A, Khan I (2010) The antinociceptive activity of *Polygonatum verticillatum* rhizomes in pain models. J Ethnopharmacol 127(2):521–527
18. Nadkarni KM (1976) Indian meteria medica, vol 1 and 2, 3rd edn. Popular Book Depot, Bombay, pp 225–270
19. Meng, QHHLYLXWSZC Key Laboratory of Weed and Rodent Biology and Management, Institute of Plant Protection, Chinese Academy of Agricultural Sciences, Beijing 100193, China). Chemical compositions and allelopathic potential of volatile oil from *Sorghum halepense* (L.) Pers. *Acta Phytophylacica Sinica*, 2009-03.
20. Saleem M, Ahmad M, Ahmad A (1998) Chemistry of the medicinal plants of Genus *Acacia*. Hamdard Med 41(1):63–67
21. Nadkarni KM (1982) Indian meteria medica, vol 2, 3rd edn. Popular Book Department, Bombay, pp 265–270
22. Sumeet D, Raghvendra D, Kushagra M (2008) *Achyranthes aspera* Linn. (Chirchira): a magic herb in folk medicine. Ethnobotan Leaflets 12:670–676
23. Nadkarni KM (1927) The Indian materia medica, vol I, 3rd edn. Popular Book Department, Bombay
24. Ashok KBS, Lakshman K, Jayaveera KN, Nandeesh R, Tripathi SNM, Krishna NV, Manjunath M, Suresh MV (2009) Estimation of rutin and quercetin in *Amaranthus viridis* L. by high performance layer chromatography (HPLC). Ethnobotan Leaflets 13:437–442
25. Rai SK, Mallavarapu GR, Rai SP, Srivastava S, Singh D, Mishra R, Kumar S (2005) Constituents of the flower oil of *Carissa opaca* growing in the Aravalli mountain range at New Delhi. Flavour Frag J 21(2):304–305
26. Tyler VE, Bradley R, Robers JE (1981) Pharmacognosy, 8th edn. K. M.Varghese Co., Bombay, p 250
27. Nadkarni KM (1976) Indian materia medica, vol 1 & 2, 3rd edn. Popular Book Department, Bombay, pp 225–270
28. Haq I, Rehman M (1990) Medicinal Plants of Upper Swat (NWFP), Pakistan. Hamdard Med 33(3):51–68
29. Ikram M, Hussain SF (1978) Compendium of medicinal plants. P.C.S.I.R. Laboratories, Peshawar, pp 23–147
30. Zaman MB, Khan MS (1970) Hundred drug plants of West Pakistan. Printing Corporation of Pakistan, Islamabad, pp 1–73
31. Versiani MA (2004) Studies in the chemical constituents of *Bombax ceiba* and *cuscuta reflexa*. University of Karachi/H.E.J. Research Institute of Chemistry, p 319
32. Rastogi RP, Mehrotra BN (1993) Compendium of Indian medicinal plants, vol 1. CDRI, Lucknow and PID, New Delhi, pp 174–178
33. Hamid S, Sabir AW, Saleem M, Chaudhry TA (1998) Medicinal plants of family Cannabaceae of Pakistan. Hamdard Med 41(1):111–112
34. Sahreen S, Khan MR, Khan RA (2009) Evaluation of antioxidant activities of various solvent extracts of *Carissa opaca* fruits. Food Chem 122(4):1205–1211
35. Rahman IU, Khan MA, Arfan M, Akhtar G, Khan L, Ahmad AU (2007) A new 14-membered cyclopeptide alkaloid from *Zizyphus oxyphylla*. Nat Prod Res 21(3):243–253
36. Bhattacharjee SK (2000) Hand book of aromatic plants, 1st edn. Popular Offset Service Pvt. Ltd., Jodhpur, pp 119–471
37. Chopra RN, Chopra IC, Honda KC, Kapur LD (1982) Chopra's indigenous drugs of India, 2nd edn. Academic Publishers, Delhi, pp 125–130
38. Zafar R, Ali SM (1998) *Cichorium intybus* Linn.: a review. Hamdard Med 41(4): 98–109

39. Ramirez I, Carabot A, Melendez P, Carmona J, Jimenez M, Patel AV, Crabb TA, Blunden G, Cary PD, Croft SL, Costa M (2003) Cissampeloflavone, a chalcone-flavone dimer from *Cissampelos pareira*. Phytochemistry 64(2):645–647
40. Gonzalez ME, Dominguez RP, Moreno DA, Garcia CV (2010) Natural bioactive compounds of *Citrus limon* for food and health. J Pharm Biomed Anal 51(2):327–345
41. Loffler C, Sahm A, Wray V, Czygan FC, Proksch P (1995) Soluble phenolic constituents from *Cuscuta reflexa* and *Cuscuta platyloba*. Biochem Syst Ecol 23(2):121–128
42. Ansari KA (2000) Recent development in the chemistry and botany of Dalbergia sissoo Roxb. (Shishm). Hamdard Med 43(2):45–50
43. Hamid S, Sabir AW, Yamin M, Chaudry TA (1996) Medicinal plants of families Salvadoraceae, Sapindaceae, and Padaliaceae of Pakistan. Hamdard Med 39(3):96–101
44. Goleniowski ME, Bongiovanni GA, Bongiovanni L, Palacio CO, Cantero JJ (2006) Medicinal plants from the "Sierra de Comechingones", Argentina. J Ethnopharmacol 107:324–341
45. Schippmann U, Leaman DJ, Cunningham AB (2002) Impact of cultivation and gathering of medicinal plants on biodiversity: global trends and issues. In: Biodiversity and the ecosystem approach in agriculture, forestry and fisheries, Ninth regular session of the commission on genetic resources for food and agriculture. FAO, Rome, pp 1–21
46. Jeong WS, Lachance PA (2001) Phytosterols and fatty acids in fig (*Ficus carica*, var. Mission) fruit and tree components. J Food Sci 66:278–281
47. Louis P, Patrick P, Andre M, Jean-Marie B, Andre F, Jean-Paul R (2000) Bergapten content in fig leaves. Ann Falsifications l'Expertise Chim Toxicol 93:427–435
48. Ahmad S, Kalhoro AM, Kapadia Z, Badar Y (1993) "Aloe", a biologically active and potential medicinal plant. Hamdard Med 36(1):109–115
49. Rahman AU, Ahmad S, Bhatti MK, Choudhary MI (1995) Alkaloidal constituents of *Fumaria indica*. Phytochemistry 40(2):593–596
50. Makkar HPS, Becker K (1998) Do tannins in leaves of trees and shrubs from African and Himalayan regions differ in level and activity? Agroforest Syst 40:59–68
51. Sabir AW, Hamid S, Chaudhri TA (1995) Medicinal plants of the family Araliaceae of Pakistan. Hamdard Med 38(4):91–93
52. Hamayun M, Afzal S, Khan MA (2006) Ethnopharmacology, indigenous collection and preservation techniques of some frequently used medicinal plants of Utror and Gabral District Swat, Pakistan. Afr J Trad CAM 3(2):57–73
53. Said M (1970) Hamdard pharmacopia of eastern medicine. Hamdard National Foundation, Pakistan
54. Dymock W, Warden CJH, Hooper D (1972) Pharmacographia indica. Kegan, Paul, Trench, Trubner & Co. Ltd, London, pp 466–476
55. Ahmad M, Sabir AW, Akhtar S, Bhatty MK (1987) Medicinal plants of family Umbelliferae (Part III). Hamdard Med 33(3):3–19
56. Haq I, Shah M (1986) Medicinal plants of Peshawar district (NWFP), Pakistan. Hamdard Med 29(3):73–87
57. Youngken HW (1950) Textbook of pharmacognosy, 6th edn. Blakistan Co., Philadelphia, pp 275–280
58. Riaz N, Malik A, Rehman A, Ahmed Z, Muhammad P, Nawaz SA, Siddiqui J, Choudhary MI (2004) Lipoxygenase inhibiting and antioxidant oligostilbene and monoterpene galactoside from *Paeonia emodi*. Phytochemistry 65(8):1129–1135
59. Rehman HU, Yasin KA, Choudhary MA, Khaliq N, Rahman AU, Choudhary MI, Malik S (2007) Studies on the chemical constituents of *Phyllanthus emblica*. Nat Prod Res 21(9):775–781
60. Shi Q, Zuo C (1992) Chemical components of the leaves of Pistacia Chinensis Bge. Zhongguo Zhong Yao Za Zhi 17(7):422–423, 446
61. Atta AH, Nasr SM, Mouneir SM (2006) Potential protective effect of some plant extracts against carbon tetrachloride–induced hepatotoxicity. Afr J Trad CAM 3(3):1–9

62. Yang ZJ, Zheng YN, Xiang L (2007) Study on chemical constituents of *Portulaca oleracea*. Zhongguo Zhong Yao Za Zhi 30(10):1248–1250
63. Chandra S, Sastry MS (1988) Phytochemical investigations on *Prunus persica* heart wood. Indian J Pharm Sci 50(6):321–322
64. Khan FW, Khan GP, Marwat AG (1977) Chemical evaluation of *Quercus ilex* seed and its oil. Pakistan J Forest 27(2):98–100
65. Gu Q, Wang RR, Zhang XM, Wang YH, Zheng YT, Zhou J, Chen JJ (2007) A new benzofuranone and anti-HIV constituents from the stems of *Rhus chinensis*. Planta Med 73(3):279–282
66. Kaul VK, Rajee K, Sing GB (1999) Volatile constituents of the essential oil of flowers of *Rosa brunonii* Lindl. Flav Frag J 14(1):9–11
67. Rafique SM (1994) Effect of grazing management and fertilizer application on vegetation and soil properties of a moist temperate forest range in Siran Valley (Mansehra), NWFP (Pakistan). Pakistan J Forest 44(4):20–29
68. Zhang LS, Li Z, Mei RQ, Liu GM, Long CL, Wang YH, Cheng YX (2009) Hastatusides A and B: two new phenolic glucosides from *Rumex hastatus*. Helv Chim Acta 92(4):774–778
69. Karick CR (1994) Pharmacopoeial standards of herbal plants, vol I. Indian Book Centre, Delhi, pp 10–58
70. Ali MS, Ahmad VU, Azhar I, Ghani KU (1998) Some medicinally important plants and their uses. Hamdard Med 41(2):96–102
71. Schutz K, Carle R, Schieber A (2006) *Taraxacum*—a review on its phytochemical and pharmacological profile. J Ethnopharm 107(3):313–323
72. Wu TS, Shi LS, Kuo SC (1999) Alkaloids and other constituents from *Tribulus terrestris*. Phytochemistry 50(8):1411–1415
73. James B, Perianayagam S, Sharma K, Pillai KK (2006) Anti-inflammatory activity of *Trichodesma indicum* root extract in experimental animals. J Ethnopharm 104(3):410–414
74. Riaz M, Ashraf CM (1990) The chemistry of indigenous medicinal plants of the genus *Vitex* (family Verbenaceae). Hamdard Med 33(2):74–83
75. Mirjalili MH, Moyano E, Bonfill M, Cusido RM, Palazon J (2009) Steroidal lactones from *Withania somnifera*, an ancient plant. Molecules 14:2373–2393
76. Das PK, Goswami S, Chinniah A. Panda N, Banerjee S, Sahu NP, Achari B. Woodfordia fruticosa: traditional uses and recent findings: a review. J Ethnopharm 21:189–199
77. Ram G, Bhan MK, Ahuja A, Meena SR, Kaul MK, Gupta KK, Jolly RL, Khajuria RK (2007) Variability and selection on different *Argyrolobium roseum* accessions for morphological traits and yield. Genet Resour Crop Evol 54:649–654
78. Turner NJ (1995) Ethnobotany today in northwestern North America. In: Schultes RE, von Reis S (eds) Ethnobotany: evolution of a discipline. Dioscorides Press, San Francisco, pp 264–283

Index

A.M. Abbasi et al., *Medicinal Plant Biodiversity of Lesser Himalayas-Pakistan*,
DOI 10.1007/978-1-4614-1575-6, © Springer Science+Business Media, LLC 2012